Dream Notes XP

Short and to the point notes and shortcuts on Microsoft's Office

by

Kirt C. Kershaw

1663 LIBERTY DRIVE, SUITE 200
BLOOMINGTON, INDIANA 47403
(800) 839-8640
WWW.AUTHORHOUSE.COM

© 2004 Kirt C. Kershaw
All Rights Reserved.

First published by AuthorHouse 09/22/04

ISBN: 1-4184-8456-3 (sc)

Printed in the United States of America
Bloomington, Indiana

This book is printed on acid-free paper.

Table of Contents

Word 52

Project 69

Access

Overview

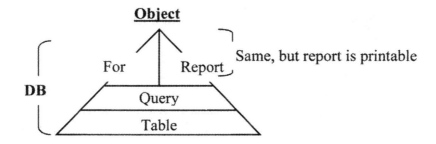

Form = Display info from table/query; or enter new info
Report = print results of forms or queries
Query = to retrieve data from a table
Table = raw organized data

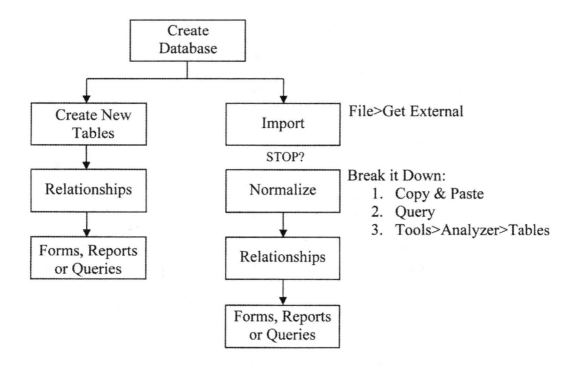

Mouse click /	To >	Task Pane <u>TP</u>:
Button (b)	Enter or Return®	

Database – The Database is a program you create to store your table(s) which in turn will store your records or your raw organized data: 1st Create your database by opening Access & in *Task Pane* /Create Database *link*, choose a place you'd like to store your data, then give it a name & /Create (b)

Tables – Are like Excel Spreadsheets in that they will have many cells for you to create & store you records in: In your Database under 'Objects' select 'Tables' & in its adjacent *window* //Create Table in Design View…

Design View: Each Table has basically 2 views one is a Design view to create titles for each field you'd like to see in your record i.e. First Name, Last Name Address etc.

Field Name: To name the field i.e. First Name, Last Name or whatever…

Data Type: Specifies the data stored in a table field;

- Text: Will store both numbers & text in this field, but no calculations can be performed for this field
- Memo: Stores more characters than the Text Data type
- Number: Use this Data Type for fields you'd like later to perform calculations with like cost, gross pay, employees hours
- AutoNumber: Is used for numbering your records, a unique identifier for each record
- Yes/No: A box will display that you can check after you change to your datasheet view that you can check like do your records have health insurance check yes or leave blank for no

Description: Here you can type in a detailed description of what each field is for & how it's to be used.

Field Properties: Used to refine what and how data can be entered into your fields: After you've typed in your field name & selected a Data Type, press F6 to toggle down to its Field Properties…

Validation Rule: To limit data being entered into a field i.e. **<30** (only values less than 30 are accepted) OR **<=30** (Less than OR equal to 30, and know that the equals sign comes ALWAYS at the left of a < or > sign or you'll get an error!)

Default Value: Enter a value you want as a default for this field every time you start a new record i.e. 40 for the average employee hired by your company; as default you won't have to type 40.

Input Mask: Symbols to define to user amount of characters needed i.e. (###) ###-#### (Like a phone #), / Input Mask's Build (b) & follow the directions

Help: Press **F1** on your keyboard when your cursor is in any field in Design view to bring up details and definitions of what that field.

Datasheet View: After you've created your fields for your data to be entered into then change your view to enter in those fields the info that will make up each record: On *toolbar* /Datasheet (b)

Entering Data: To save a record, move into another record or hold SHIFT+®. If stuck with data entry, always press ESC once or twice. If you need to make Design changes to your table on *toolbar* /Design View (b)

Filter By Selection: To do quick filter by what you have selected i.e. you're in the 'State' column & in the field that contains UT for Utah & you want to only see all records containing UT: Right / that UT *cell*>Filter By Selection. To remove your filter: Right / anywhere in the table>Remove Filter Sort

Filter Excluding Selection: To filter OUT or remove all fields that you've selected. If your in the 'State' column for your records & you want to view all records excluding NY (New York) Right / the cell that contains NY in *any* record>Filter Excluding Selection. To remove you filter: Right / anywhere in the table>Remove Filter Sort

Remove Filter Sort: After, Sorting, Filtering by either Selection or Excluding, to revert your table back to its original view: Right / any cell>Remove Filter/Sort

Backup: to back up a table: Right / table desired to be backed up>Copy, then Right / white space in Database>Paste & select & type new name for the Table & select a type…

- Structure Only – paste's only the fields of a copied table
- Structure & Data – paste's the fields & their data into a new table
- Append Data to Existing Table – here you type in the name of an existing table you'd like to paste (append) the data you've copied from the other table. (The field names need to be the same when appending too)

Relationships – Getting multiple tables to interact by hooking up similar values & Data Type fields between tables. Putting all of your information into one table is inefficient and slows down your processing speed.

Build: *toolbar* find and /Relationships and add whatever tables to the gray field that you want to form relationships. Then / & drag one field name from one table to another (both fields must have same Data Type & Properties); after hooking them up you'll have window pop-up with a few options to check:

- ✓ Enforce Referential Integrity: Check this box, because without it you could create "Orphan" *records*. Orphans are records that don't have a corresponding record in the 'Initiator' or 'Primary' table, and hence will produce inaccurate results every time you query a table that contains Orphan records. But checking this box alone will prevent you from adding, deleting or changing ANY records including ALL tables it's related to. So check also the Cascade options…
- ✓ Cascade: Check the 2 Cascade options, and they will allow you to later update & modify records in related tables. But when making changes ALWAYS start with the 'Initiator' table 1st. By also starting with the 'Initiator' first, its corresponding linked key fields with other tables will update themselves automatically too.

Edit: To Edit or delete a relationship // the thin line connecting the related tables & make your changes

1 to 1: After you create a relationship and notice it's a one-to-one relationship, it's important to remember the 1st table you dragged your field from (the Initiator of the relationship) that created the relationship with the other table, because later when you want to add, delete or make any changes with the records in the two tables, Access only allows changes to be made in the 'Initiator' table FIRST. After, you can add, delete or change records in the 2nd related table.

Mouse click /	To >	Task Pane TP:
Button (b)	Enter or Return®	

1 to Many: If your relationship between tables is a one to many then it's important later when you want to add, delete or make any changes with the records in the tables to start with the table that is labeled the '1' relationship & then after you can make changes with its related 'Many' table.

Subdatasheets: If you're in the Datasheet view of your Table & prior you've created a relationship with this table to another, you can view that other table within this table you're viewing as a Subdatasheet: Insert>Subdatasheet (By default your related table will be highlighted in blue, /ok to accept. If you have more than one table related to this one then make your selection of the other related table & /ok). After you /ok, at the beginning of each record you'll see a + sign, / it to expand that record's to view its related record from the other related table.

Remove: Format>Subdatasheet>Remove

Query – If you are looking for a particular record(s) and don't have the time to scroll through your table of 100's or even 1000's of records you can use the Query to get precise results out of your tables: Under Objects /Query (b), // Create Query in Design View, a list of all your tables will pop-up so select one & / Add (b), /Close (b). Your table you've added is inserted into the Query with a list of ALL its fields & is now called a 'Field List'…

Field List: From your field list find the fields you want to see in your query & //each field & it will add it automatically below, each in its own 'Field' cell…

Criteria: Next type in the Criteria cell of your corresponding field(s) the text you want an exact match for: >For example, if I wanted to pull only those records that are from Utah and I had the State field from my Field List added to the grid below, I would enter in its Criteria cell UT (or Utah, depending when you created you table if you had is spelled out or not). When finished on *toolbar* /Run (b) or the Datasheet (b) to see results. *Toolbar* /Design View (b) to go back & edit your query.

Advanced: Advanced Criteria you can use in your queries:
- **<=30** (Desired result to be less than or equal to 30)
- **>=40** (Greater than or equal to 40)
- **30< And <40** (#'s greater than 30 &less than 40)
- **UT Or ID** (To see only Utah & Idaho clients)
- **<4/12/03** (All dates older than April 12, 2003)
- **m*** (The asterisk is a 'wild card.' Anything that stars with the letter M)
- **Is Null** (Looking for fields that have no values)

Or: Anything in the "or" field is not bound by *any* Criteria: For example, if you did a query with the >criteria's – **UT & >$50,000** (All Utah client's earning more than $50 thousand, & you decided you'd like to include in your criteria Idaho. Type **ID** in the 'or' field & it will pull up ALL Idaho clients irregardless of your $50,000 criteria.

Edit: After you've ran your query you can make changes to those records in the query & it will automatically update those records in your table too.

Calculating: Right / a blank Field *cell*>Build, / = (b), and in the middle column, // the field you want to add to the equation, / * (b) (asterisk to multiply), // your 2nd field Name you want multiplied with the first field & /ok i.e. = **[WeeklyHours]*[HourlyRate]**. To change the Name of your new field selecting in that cell 'Expr 1' and replace it with **Weekly Gross**.

Forms – Forms are another way creating or editing records in your tables: Under Objects /Forms, //Create Form by using Wizard, select a table and />> *arrows* (b) to move ALL available data over (you can then select another table & />> again to add more data from another table to your form) & follow the rest of wizard to complete your form. After you form is completed you'll be in the Form *view*...

Form View: In this view you can view your records from the table(s) this form was based from when created OR enter (or edit) new records, & they will automatically be added (or updated) to the table the form was based on: Bottom, lower-left of your form are some black triangles.

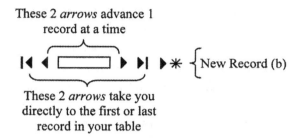

These 2 *arrows* advance 1 record at a time

New Record (b)

These 2 *arrows* take you directly to the first or last record in your table

Design View: You can get to this view from your Form *view* up on *toolbar* /Design View (b) (To get back to Form view /Form View (b)), In this view just like in the Design view of your table you can make behind-the-scene changes to enhance the value & look of your form for other users including yourself...

Label: A Label does nothing more than places itself as a label next to a field (Called a Text Box) in your Form. To edit or change the name of your Label: //the Label & type your new text in.

Text Box: A box that displays the information from the fields of each record of the table your form was based off of i.e. Names, Addresses etc.

Moving: To move a Label with its Text Box: / on either one and hover your mouse over it until you see a *hand* them / & drag the hand. For just moving the Label or Text Box individually: / on Label (or Text Box) and hover over upper-left corner of it till you see a *pointing finger*, then / & drag the finger.

Field List: The field list is a floating box which lists all the fields from the table the form was based on. If you accidentally delete a field on your form you can drag these names from the Field List onto the form (Note: if you don't have a field list that means either you haven't built your form using the wizard based off of your tables, or the Field List is hiding in which case up on the toolbar /Field List (b))

Toolbox: This floating toolbar is used to help beautify & help you setup to perform some calculations based of your # fields. If you don't see your Toolbox, on your *toolbar* /Toolbox (b)

Label: Using the Label option you can add boxes with text in it; like adding Form Header title: Hover your pointer between the two bars titled 'Form Header' & 'Details' (If you don't have a 'Form Header' bar then View>Form Header Footer), & / and drag down to pull the 'Detail' *bar* in order to

Mouse click / Button (b)	To > Enter or Return®	Task Pane <u>TP</u>:

create spacing between the 2. On Toolbox /Label (b), then / & drag a box below the 'Form Header' bar & type your text in. After you type your text in hit ® & then you'll have access to the Formatting buttons on the Formatting *toolbar*

Text Box: You can add your own blank Text Box to the form in which you can program it (With the help of the Property Sheet) to add, multiply, subtract or divide etc., other # fields in your form: On Toolbox /Text Box (b) then find a place on the Form grid & / to add your Text Box (Also a Label Box will be added to later Label your Text Box)…

Properties: The Property sheet is used to customize the particulars of your Form including Labels & Text Boxes: To finish programming your blank Text Box Right / it > Properties, /All *tab*, / in Control Source *field* & you'll see 3 dots at its end (called Build Button), /Build (b), press = (b), and in the middle column //the field you want to add to the equation, press * (multiply), .//on 2nd field Name you want multiplied with the first field & /ok. Below the Control Source *field* is the Format *field* / in it & then /its *arrow*>Currency (so our formula will be displayed in $) i.e. **= [WeeklyH ours]*[HourlyRate]**. Name its new adjacent Label box by triple clicking on it and type in **Weekly Gross** & hit ®.

Reports – Reports have the exact same Design views as Forms, but different data output views. Basically, everything is the same executing & setting up your Report as you did above when you creating your Form (except of course you'd //Create Report using Wizard)

Print Preview: This is the displayed report in printable format based upon the table or query you selected when you created the report using the Wizard.

Design View: This is exactly the same view as in the Form's Design View (follow the steps in Form's Design view to Design, edit or manipulate your Report)

Report Header: To display a title at the beginning of your 14 page Report: View>Report Header Footer, On Toolbox /Label (b), then / & drag a box below the 'Report Header' bar & type your text in. After you type your text in hit ® & then you'll have access to the Formatting buttons on the Formatting *toolbar*

Page Header: To display a title at the top of each page: View>Page Header Footer & add a Label Box under the Page Header bar

AutoFormat: To change Reports appearance to another formatting template: 1st select the report Edit>Select Report, on *toolbar* /AutoFormat (b) & select another template

Import Excel – File>Get External Data>Import, browse & change your File Type>Microsoft Excel, find & //your Excel file & follow the wizard.

Export to Excel – File>Export & change your File Type>Microsoft Excel 97-XP & pick a place to export your table to & /Export (b)

Query Joins – Displaying two or more tables in a query and creating temporary joins, by dragging one field from one table to another field with the same value in another table (just like creating relationships). By default created joins are Inner Joins. Warning! 2 or more tables not joined in a query will always give you bogus data. (Bonus Note: If you get an error from a // to open a query, you must Right / the query>Design View).

Inner Join: Displays only records from "both" tables that have matching values in the "joined fields"

Outer Join: To display all records from one table & only the matching records of the 2nd table i.e.

After you have created a join, // that thin line (join) between the 2 tables and select #'s 2 or 3 according to their explanations and your desired results.

Self Join: To join a table to itself i.e. if you have an employee table that includes a column for supervisors, >and you want to display each employee with his supervisor: Display the same Employee table twice and join from one table's *employee id* to its duplicate table's *supervisor id* (Note: you can't match a field from 1st table to same field of 2nd table, or id to id, it would be like a snake eating its tail and you'd get errors. *Supervisor id* has same value as Employee id but can be joined because they're separate fields in the table). Then add the Employee Id from 1ˢᵗ table & Supervisor Id from 2ⁿᵈ table to the grid below & Run Query.

3 Table Join: If you want to join two tables but they have no matching fields & values, you can bring in a 3rd table that can link or bridge the two through its matching fields with both tables.

Top Value Query – To query top 5 of anything i.e. sales people or products sold: *Toolbar* /Totals (b) (adds the Totals field to the grid below but as a 'Group' for each field in the grid. Remember you can only have one 'Group' so all the others have to be changed>Sum or deleted from the Query). On *toolbar* /Top Values (Currently displaying the word "All") *arrow*>5 & Run Query

Parameter Query – Parameter Query's are nothing more than a prompt to the user (you) to type in an answer that will be used as criteria in your query after you run it. For Example, if you only want to view clients from a particular state like Utah, in your query: Add that State *field* to the grid below & in its criteria cell type **[Enter a State]** Run your query & when prompted type in Utah, /ok. Other Parameter Queries include…

★ Wildcards: Asterisks * used to abbreviate inquiries i.e. type in your criteria cell for your State field **Like [Enter the first letter of a State]&*** (the ampersand links parameter values to wildcard). So after you type in U for 'Utah' & /ok your query will pull up all those states that begin with the letter 'U'

★ Between: Between is used in a double parameter query to pull up #s or dates data between other #s or dates i.e. in your criteria cell of a Date *field* & type **Between [Enter a start date:] and [Enter an end date:]** (when you run your query this will ask the user the 2 questions above & as a result pull all the dates between what they entered in the start and entered in the end dates).

Update Query – To update specified field(s) in your table using the Update Query. For example, let's say we sell Licorice & it's time to update our prices by 5%: On *toolbar* / the *arrow* to the right of Query Type>Update, this puts an update field down in the grid, & in that field type in **[Licorice]*1.05** & /Run & /Yes. (Note: If you run & Update, Append or a Delete Query on a table that you have open, the results won't show after your run the queries unless you close that table & reopen it OR Right / anywhere in your table>Remove Filter Sort)

Mouse click /	To >	Task Pane TP:
Button (b)	Enter or Return®	

Append Query – Used to add records from one table to another (Note: You must have matching field names & data type when appending from 1 table to another. If some fields don't match, then access will ignore them and only append those that do): On *toolbar* / the *arrow* to the right of Query Type> Append, /Table Name *arrow*>a table you want the records added to & /ok, /Run, /Yes

Delete Query – To delete records from a table. First run your query & /Run to view the records you'd like to delete & then: /Query Type arrow>Delete Query, /Run, /Yes.

Advanced Forms – To add more functionality to your forms go to its Design *view* and…

    Pictures: To insert and picture into your form: From floating *Toolbox*, /Image (b) & /on grid wherever you want your image, then browse & select your image

    Combo Box: Creates a drop down list of fields from your table: On *toolbox*, (make sure Control Wizards (b) is selected or highlighted) & /Combo Box (b), & / on grid wherever you want your drop down list to go, /Next, select a table, /Next & select all the fields you want, /Next & uncheck the Hide *box*, /Next, /Next & select Store That Value *arrow*>a key *field* (like a product field if your main purpose if using a product # field as a lookup), /Next, /Finish. Go to your Form *view* & test it!

    Options: To add radial dials or options to your form. For example, you can have options for payments i.e. >Credit Card, Check etc: On *Toolbox* (make sure Control Wizards (b) is selected or highlighted), /Option Group (b) & /on blank area of your form & type in the *labels*: Credit Card (press tab) Money Order (press tab) etc., /Next, /Next & select a store value field (like a Payment field) and go through the rest of the wizard; then finish & test in Form *view*!

    Command Button: To create a (b) on your form & assigning a function to it i.e. closing the form the button currently resides in or opening another form: In floating *Toolbox*, /Command (b) & / anywhere on the form, & in left *window* pane select 'Form Operations' & in right pane select 'Close Form', /Next & select 'Text' option, /Next & title it 'Close Form,' /Finish

    Subform: Inserting one form into another .But 1st make sure forms are linked so all records displayed in subform are related to current record displayed in Main form, and it's nice too so you can view a record in one form & see its corresponding record in the subform (Make sure Control Wizards (b) is selected or highlighted): On *toolbox* /Subform (b), then / anywhere on the form: /Next & select a table data source for your subform & add desired fields, /Next, & type in Name of your Subform, /Finish & Test it!

Advanced Reports – To add more functionality to your Reports go to your Design view and…

    Hiding Duplicates: Hiding repeating info in a form. For example, if you have a customer listed in your report several times because they've ordered more than 1 product, but you only one your customer listed once with ALL of their products ordered: Right / on the Text Box that you want not repeated on your form>Properties, /All *tab*, find & //the Hide Duplicates *field*>Yes.

    Grouping: To keep groups of info together. For example, let's say in your report you have a list of DVDs you sell & each has a field in a table that you've put into a category of style i.e. mystery, comedy etc. You can group ALL by 'Style' on your report: On *toolbar*, /Sorting & Grouping (b), & / in a blank cell under Field Expression, then /its *arrow*>Style, below /Keep Together *arrow*>Whole Group & Close out of Sort & Group & test!

<u>Sum Fields</u>: To add up fields. For example, we have to field in our table of 'Instock' & 'Onorder;' & so we can add the two field to get our Total Inventory: 1ˢᵗ on *Toolbox* /Text Box (b) & name its label **Total Inventory** (/// label to rename it), and Right / the Unbound Text Box> Properties, /in Control Source *field* & type **=([instock]+[onorder])**

<u>Mailing Labels</u>: Create mailing labels from your table using wizard: Under Object select Report & above objects is the Database *toolbar*, /New (b) & select Label Wizard & follow wizard's steps.

Reset AutoNumber – When you records in a Table uniquely defined by AutoNumber & delete one, AutoNumber does not automatically renumber itself & you're left with # gaps i.e. 1, 2, 4, 5, 9, 10 etc. The following steps will show you how to reset AutoNumber for a table that has one reference. If there is more than one table referenced, you'll have to repeat these steps for each referenced table:

1. Delete the relationships between tables.
2. Main Table:
   a. Change the 'Data Type' of your AutoNumber field in your main table to Number & then remove the Primary Key, /Save (you MUST /Save before proceeding!)
   b. Create a new AutoNumber field & name it 'Main Table Field', /Save, & close the table.
3. Referenced Table:
   a. Open up your table that your main table will be referenced to & create a new Number *field* & Title it 'Referenced Table Field,' & save it, & close the table.
4. Update Query: We need to create an Update Query that updates the new field in your referenced table to the new AutoNumber *field* in your main table
   a. Create a query in Design view
   b. Add your two tables, Main & Referenced
   c. Find the field in your Main table that was previously linked to the Reference table & drag to re-link those fields together to create a join.
   d. On toolbar / the *arrow* to the right of the Query Type (b)>Update Query
      i. //Referenced *field* titled 'Referenced Table Field' (to add it to grid below)
      ii. In its Update To *cell* type 2 sets of brackets: in the first set will be the name of your main table, period, and in 2ⁿᵈ pair of brackets the name of your new AutoNumber *field* of your main table i.e. if you main table's name was 'Candy' & the new AutoNumber field in that table is titled 'Main Table Field'…**[Candy].[Main Table]** (this syntax will update the new field values in the referenced table).
      iii. Run the Query & /Yes to update # row(s)
      iv. File>Close & /No to save.
5. Delete the original linking field from your main table & the referenced table.

Mouse click / Button (b)	To > Enter or Return®	Task Pane <u>TP</u>:

6. Rename the *new* AutoNumber field to original name
7. Re-create the primary key & the relationship between tables.

Table Analyzer – helps to create new tables from an improperly designed one that already contains data. The Analyzer will eliminate repeating info in same fields by placing them in 2 or more related tables & can only be run on one table at a time: Select your table & Tools>Analyze>Table, /Next, /Next, /Next, /Next & move tables around so you can see them, if you like & agree with the 'Analyzer' then rename the tables: select a table & /Rename Table (b) & type in a name, /ok & finish naming the rest, /Next, /Next, /Next, /Yes & make sure 'No, don't create query' is selected & uncheck 'Display help…', /Finish, /ok.

Import Objects – to import objects from other databases i.e. tables, forms etc: File>Get External Data>Import, find & //your database you wish to import from, then select tables, forms etc. to import & /ok

Many-To-Many Relationships – Access does not support them. It's best to create an intermediary table that contains both primary keys of the 2 tables with the many-to-many relationship; and hence to have the 3rd table create one-to-many relationships with the other 2 tables:
1. Create 3rd Table: create a new 3rd table containing the exact same names, data types & field properties of what you consider to be the primary key fields in the other 2 tables. Make those 2 fields in your new table both Primary Keys & Save & close the table
2. Add Records to 3rd Table: Next, you'll want to copy (append) the records from one of the 2 tables to your 3rd Table: //Create Query in Design & add 1 of the 2 tables involved in the many-to-many relationship, well call it Table 1. Add what you would consider to be the primary key field & the foreign key field (the field that is linked to table 2's primary key field) to the Query grid below. On *toolbar* /the *arrow* to the right of the Query Type (b) >Append Query, & /Table Name *arrow*>your 3rd table, /Close, /Run, /Yes, (Error: because Table 1 has duplicates, which is a violation to append to Table 3 because of the Primary keys that won't allow duplicates. So some records won't append), /Yes & Close Query without saving.
3. Restructuring Tables 1 & 2: Next, to restructure the 2 tables into 2 new tables so as to comply with the 3rd intermediary table's design – removing duplicate fields from the 2 tables: Right / Table 1>Copy, Right / a blank area in your Database>Paste & type new temporary name for the restructuring of Table 1 i.e. Table1Temp & select 'Structure Only,' /ok…
   - Structure Only – paste's only the fields of a copied table
   - Structure & Data – paste's the fields & their data into a new table
   - Append Data to Existing Table – here you type in the name of an existing table you'd like to paste (append) the data you've copied from the other table. (The field names need to be the same when appending too)
Right / Table 2>Copy, Right / white blank space in your Database>Paste & type Table2Temp & select 'Structure Only,' /ok
4. Remove Foreign Key Fields: Now you'll want to delete the fields you copied from the TempTables 1 & 2 that were copied & joined & creating the annoying many-to-many

relationships from the original tables. Also, find what you consider to be Primary key fields in your TempTables 1 & 2 & make them Primary Key fields.

5. <u>From Original>TempTables</u>: Next, is to copy the records out of the original tables & paste them into the TempTables 1 & 2 using Append Query: //Create Query in Design, Add Table 1 to Query. Add all the fields from Table 1 into the Query grid, /Query Type *arrow*>Append Query, /Table Name *arrow*>Temp1Table, /ok, /Run, /Yes, /Yes, Close & query & don't save. Repeat the same steps to add all records from Table 2>Table2Temp

6. <u>Delete Old & Rename New Tables</u>: Now that you've appended all the records from the old tables & into your 2 new TempTables, you can delete the original tables. After original tables are deleted, Right / each TempTable>Rename & name them as the previously, original tables were named

7. <u>Relationships</u>: /Relationships (b) & Add Tables 1, 2 & 3 & put 3 in the middle & appropriately create the relationships.

Compact & Repair – deleted tables leave empty disk space that fragments (pigs up your hard drive space) & slows database processing time: Open the database that you've deleted any tables from & Tools>Database Utilities>Compact And Repair Database.

Crosstab Query – performs a summary calculation according to the values in row & column headings. Disclaimer: the following steps are general steps & are concepts because of the nature of what you choose tables or queries & the fields you choose from them: On Database's *toolbar* /New (b), select 'Crosstab Query Wiz,' /ok, select a table or query, /Next & select your fields, (you can specify here multiple row headings that will appear in columns at the left side of the crosstab), /Next (here you can only specify one that will appear at the top of each column) & select a value, /Next & select "Sum," (this can summarize the values in the rows & columns & are displayed in body of crosstab) /Next & /Finish

Pivot Table – to pivot between data based on queries: Open your query & View>Pivot Table View & drag the fields from list onto desired Fields on table. To add more than 1 field to Detail Data field: select multiple fields from field list holding Shift *key* & in field list at bottom /its Row Area *arrow*>Detail Data, & /Add to (b).

<u>Create New Field</u>: To create a new field that multiplies, divides etc. two fields: On *toolbar* /Calculated Totals & Fields (b)>Create Calculated Detail Field & in Name box type the name of your table & in *window* replace '0' with you syntax i.e. **hours*hourlyrate**, /Change (b), /Format *tab*, & /Number *arrow*>Currency

<u>Total Sum</u>: to get total sum of each field listed in your Rows Fields: /on your 1st column *heading* of your Detail Fields & on *toolbar* /AutoCalc (b) Σ>Sum

<u>Details</u>: to hide details from your Pivot Table: On toolbar /Hide Details (b). To see sub- detail fields & remove their parent details: In Detail Fields / + sign left of 1st column *heading* to expand. Once expanded you can see the sub-detail

Mouse click / Button (b)	To > Enter or Return®	Task Pane <u>TP</u>:

for that column you can / & drag its parent off on to toolbars above to remove it

Pivot Chart – to pivot between data based on a query in a chart: Open your query & View>Pivot Chart View & drag the fields from list onto desired Fields on chart. After, you can change the structure of your chart by switching fields around by dragging them, or to remove them drag them off of chart

Sort: to sort data ascending or descending: Right / a Datafield's (b)>Sort>Descending
Chart Type: to change chart type i.e. pie, bar etc.: Right / gray area>Chart Type: & select another type of chart

Form Pivot Chart – to create a pivot chart in a form based on a query: //Create Form by Using Wiz, & chose your query & add your fields, /Next & select Pivot Chart, /Next & select a style, /Next & type a name, /Finish & drag the fields from list onto desired Fields on chart. Next, open up a form in (Design View) you'd like your Pivot Chart inserted as subform & on *toolbox* /Subform Subreport (b) & / in a blank area of grid & select 'Use an existing form,' & select your PivotChart Form you just created & /Next & select 'Define My Own' & /Form Report *arrow*>a field you'd like linked to your chart so when you toggle through your records the chart will reflect those records, then /Subform Subreport *arrow*>same field name as you just selected so they're matching & will link your form to your sub-form (Chart), /Next & /Finish

Axis Titles: to change these title names: Right / an 'Axis Title'>Properties, /Format *tab* in Caption *box* replace text with a new name for Axis & close the box. Or you can delete these by selecting & hit Delete *key*

Macros – are commands you create that can quicken your productivity

Open Form: to create a macro to open a form: Under Objects /Macros & on Database *toolbar* /New (b), & /Action Cell's *arrow*>Echo, press F6 (to toggle to Echo's Properties) & /Echo>No (Echo No – won't show macro running in background, only its end results), /In 2nd Action Cells *arrow*>OpenForm, press F6, /Form Name's *arrow*>a form you'd like this macro to open, /Save & close Macro. To create a button for your macro…

Add Button: to apply your 'Open Form' macro to a (b) in a form: Open your form in Design view you'd like to create a (b) that will open up another form. On *toolbox* turn off 'Wizard' with a /, & /Command (b) & add it to your form. Right / your new (b)>Properties & in Name Field replace text with & name your new (b), then /Event *tab*, in 'On Click' *box* /its *arrow*>your mcrOpenForm macro, then /its Build (b) … (which will open up the mcrOpenForm macro), /in OpenForm *cell*, press F6 & /in 'Where Condition' *box* (this condition compares record in current form & finds its match of that record in the 2nd form after somebody hits your OpenForm (b)), Shift+F2 (to zoom) & type [Field Name that is the same field in the 2nd form]=[Forms]![Name of your 2nd form]![Same Field as the field in your original form], i.e. **[Clien tID]=[Forms]![frmClientData]![ClientID]** /Save & close Macro & test your (b)

<u>Require Data Entry</u>: checking the data entered & executing a macro i.e. if a field is left blank the record won't save & it will take you back to that blank field too: Open your form inDesign View, Right / dark gray area>Properties (Form Properties), /Event *tab*, /in 'Before Update' *box* (This box is used to specify events that will occur before the table is saved) & /its Build (b) …, select 'Macro Builder,' /ok & type name your macro, /ok. On *toolbar* Conditions (b) & in 2nd Condition *cell* type the Is Null expression of what field can't be blank i.e. **[ProductName] is Null** ® (this condition will run the macro if no value is entered in ProductName *field*) & / 2nd Action cell's *arrow*>CancelEvent ® In 3rd Condition *cell* type ... (Note: entering … or ellipse allows you to attach more than one condition) ® /Action cell>MsgBox ® press F6 & type **You must enter a value in the Product Name field** ®® & /Type *arrow*>Information ® **Required Data** In 4th Condition *cell* type ... ® & /4th Action cell's *arrow*>GoToControl ® & press F6 & type **[ProductName]** (In other words, when the record is not save because Product Name field was skipped this will Go back to Product Name field, Go To Control) ® & press F6 & type /Save & Close Macro

<u>Automate Macro</u> – A macro that enters a value in one field based upon a value in another field i.e. if someone enters the product name 'Gummy Bears' then in another field you called, Gifts, you could have displayed 'Free Shipping': Open your form in Design View, Right / Product text box>Properties, /Event *tab*, /in 'On Exit' *field* & /its Build (b) … & select 'Macro Builder,' /ok & type a name for your Macro, /ok, *toolbar* /Conditions (b) & /in 2nd Condition *cell* type **[Product] in ("Gummy Bears","Gummy Worms")** (this will compare values in this Product field with values in parenthesis) ® & / 2nd Action *arrow*> SetValue ® & type, press F6 & type **[Gifts]** (this is name of field), ®, & type **"Free Shipping"** (is the value you want displayed in the Gifts field). /in 3rd Condition *cell* & type ... (ellipse) ®, & / Action *arrow*>**GoToControl**, ®, press F6 type **[Address]** (specifies the control where you want the insertion point to move to, the Address *field*), /Save & Close Macro& test your macro in your form

Form Operations – the following will help customize your forms

<u>Conditional Formatting</u>: to apply formatting to text when it meets a specified criteria i.e. We can program the input of 'Quantity' to be displayed in red as a reminder to the data entry person that there's free shipping to those who order 5 of a product: Open your form in Design View, select 'Quantity' Text *box* & Format>Conditional Formatting, /2nd *arrow*>Greater than or equal to, 3rd *box* type **5**, & select a Font *color*> Red, /ok. You can also format non-number fields that are associated with your 'Quantity' Text box like the 'Product' Text box: Go to Design View of your form: /Product Text *box*, Format>Conditional Formatting, /Condition 1 *arrow*>Expression Is & in blank box type in your expression that will extract the # 'Data Type' from the 'Product' Text box field of a subform, to the 'Quantity' Text box & there add your criteria too, i.e. **Forms![frmSomeForm]![fsubSubForm]![Quantity]>=5**

Mouse click /	To >	Task Pane <u>TP</u>:
Button (b)	Enter or Return®	

& select same color you chose earlier for your 'Quantity' Text box, /ok & / View (b), to see results...

<u>Calendar</u>: to create a calendar of a form for a reference & placing a command (b) on another form to / on to open your Calendar:

1. <u>Create the Calendar</u>: //Create form in Design view, On floating *Toolbox*, /the last (b) 'More Controls'>Calendar Control 10, & grid / & drag a 4" wide, 3" high box Right / dark gray>Properties (Form Properties), & /Format *tab*, & in 'Caption' type name of you calendar, //Scroll Bars *field*>Neither, //Record Selector>No, //Navigation Button>No, //Dividing Lines>No, /Save, /Other *tab*, //MonthLength>English (Months are now full & not abbreviated), //ValueIsNull>Yes (removes today's date marker), /Save, /Form's Restore (b) /View (b) & /Save & Close your calendar form

2. <u>Create Button</u>: to create a (b) on a form with a macro that when you /on (b) will open up your other Calendar form: Open up a form in Design View you'd like your to put a (b) that will open up your calendar. *Toolbox* turn Wizard on, then /Command Button (b) & /anywhere on your form, under Categories' select 'Form Operations' & under 'Actions' select 'Open Form,' /Next & select 'the name of your calendar form,' /Next & select 'Text' & type in name box **Show Calendar**, /Next & type a name for your command (b) i.e. **cmdCalendar**

3. <u>Position Macro</u>: to create a macro that will open your calendar 4" to right of your screen: Start a new Macro, /1ˢᵗ Action cell, /its *arrow*>OpenForm, press F6 & /Form Name's *arrow*>the name of your calendar form. 2ⁿᵈ Action cell's /*arrow*>MoveSize, press F6 & type **4®** /Save & close your macro. Open your form in Design View that has your calendar (b), & Right / Show Calendar (b)>Properties, /Event *tab*, /On Click *arrow*>the name of your macro you just saved, /Restore (b) for your form & size it to occupy only half of your screen & /Save /View (b), /Show Calendar (b)... (you may have to resize your form & Calendar to view both in a small 17" monitor window)

   <u>Default Date</u>: Extra: If you wanted to view a record based upon (table or query field) a date selection the user makes from the calendar (like a project date), know that the "Value Is Null" field set to No won't work to keep calendar when opened defaulting to today's date Open Calendar Form Bring up properties for that Form & /Event *tab*, /On Load *arrow*>Event Procedure & /its Build (b), & add the following line of code **Me.Calendar.Value = Date** ...(with the "Calendar" replaced by whatever name your calendar has)

<u>Form Tab Pages</u>: organizing large forms by adding tab pages, categories: Open your form that you'd like organized into categories & in *Toolbox* /Tab Control (b) & /on the form. Right / Page 1 *tab*>Properties, On Property sheet /Other *tab* & in Name *box* type in a name for your tab. From Field List add what you want in that first tab page either from your field list OR if you already have the fields on the form you can cut & paste those into the 1ˢᵗ tab page.

<u>Add or Delete Pages</u>: simply Right / the tabs>delete or insert.

<u>Same Label Different Tab</u>: If you add the same label with its Text box on more than one tab page you'll have to rename all their labels except the 1st one, if not you'll have to use Access's confusing default labels

Tab Order: to move your tabs around so they'll be in the order you'd like them: Right / border of Tab Control *box* (the box on your grid that holds all the tab pages)>Page Order

Tabs or Buttons: to change your tabs of your tab pages into buttons: Right / border of Tab Control box>Properties, /Format *tab*, & //Style>Buttons

Power Reports – enhance your reports with the following…

Auto Cancel Report: creating a macro to prevent opening a report if the user enters invalid data in the prompts (prompts are parameters created in a query, & of course the form is based on that query that then prompts the user for criteria every time the user opens that report): Right / your Report> Design View (this prevents the report based upon your query from prompting you), Right / dark gray area>Properties (Report Properties), /Event *tab*, /On 'No Data's Build' (b) & select 'Macro Builder,' /ok & type name your macro, /ok. /1st Action cell's *arrow*> MsgBox & press F6 & type a message like "There are no records matching your criteria" ® ® /Type *arrow*>Information ® & you can type here your company's name. /2nd Action *arrow*>CancelEvent & /Save close macro. Open your report & when prompted invalid info to test it…

Chart: to insert a chart (its data based upon a query or a table) into the report 1st you'll obviously have to create your query you'd like to see its resultant data in a chart & then do the following: Open your report in Design View & view>Header & Footer>Report Header Footer (to add a report footer if you don't have one). Insert>Chart & drag your chart in Report Footer about 3" wide by 2"tall, select queries & select the name of your query, /Next & add your fields, /Next & select a chart, /Next, /Preview Chart (b) (to see if you like what you see) & /Close, /Next & select 'No Fields' for both Report & Chart Fields, /Next & type a name, /Finish

Report Columns: to shorten a 1 column report by converting it into 3 or more. For example, if you have a report that has a list of all your friends & their phone #'s you can put them into 3 columns on a couple of pages instead of 1 column stretching across several pages : Open your report in Design View, Add a Text Box to Detail *section*, delete its label, Right /Unbound text box>Properties, In Control Source box type **=[LastName]& ", " &[Phone]** (Note: the ampersands concatenate or join both fields into one) *Toolbar*, /Sorting & Grouping (b) & in 1st Field Expression *cell* type **=Left([LastName],1)** ® (The 'Left Function': will group & sort 1st letter of each Last Name), copy this expression to clipboard, //Group Header *field*>Yes, //Keep Together *field*>Whole Group & Close the box Add a Text Box to the new 'Group Header' *section* & delete its label. Right / the Unbound Text Box>Properties & Paste the Left Function in Control Source *field* (This copy of Left Function will actually ADD the 1st letter of each last name to report, remember we 1st added this function to sorting & grouping so after letter is added, it will group all last names beginning with the letter 'A', then 'B' etc.) File>Page Setup, /Columns *tab* & set # of columns>**3** & spacing>**.2**, uncheck 'Same as Detail' & in Width type **2** & select "Down, then Across", /ok. /Save & view your report

Mouse click /	To >	Task Pane TP:
Button (b)	Enter or Return®	

<u>Snapshot Report</u>: to send a report electronically to anybody including those who don't have Access, (pun intended): Under object select a Report & File>Export, & name it & / Save as type *arrow*>Snapshot Format, /Export. Look in the folder you exported it to & it will be there.

Merge with Word – to use 'Office Links' to share data from Access with Word when doing a Mail merge: Under 'Objects' select your table or query you'd like that data to be used in Word's Mail Merge & Tools>Office Links>Merge with Word, /ok, browse & you're your form letter & //it. TP, /Next & / so your cursor is on 2nd line of page & on Mail merge *toolbar* /Insert Merge Field (b) & //the fields you'd like to add to your form letter & close field list when finished. TP, /Next, /Next & /Print *link* to choose what to print…

Publish to Word – to publish your Report to Word: Under object select your report & Tools>Office Links>Publish to Word

Publish to Excel – to publish a table or query to Excel to use special charting features Access doesn't have:
Under Objects select your query & Tools>Office Links>Excel

Switchboard – is a form (or one big fat menu) created to help the user navigate their database easier i.e. if you have 3 forms that the user of your database will be operating with, create a switchboard or one main form for the user to have easy access to all these forms (Note: Once you created your switchboards, Access creates a Switchboard table; so if you want to modify a Switchboard make the changes in its Table: Open Switchboard Item table): Tools>Database Utilities>Switchboard Manager & select 'Main Switchboard' & /Edit, /New & type the **Name of &Form** (placing an ampersand before any letter creates a shortcut key, and in this case Alt+F), /Command *arrow*>Open Form in Add Mode, /Form *arrow*> choose your form, /ok
1. /New & type **Name of &bForm**, (shortcut key, Alt+b), /Command *arrow*> Open Form in Add Mode, /Form *arrow*>choose your form, /ok.
2. /New & type **Name of &cForm**, (shortcut key, Alt+c), /Command *arrow*> Open Form in Add Mode, /Form *arrow*>choose your form, /ok
3. /New & type **Close &Application**, (shortcut key, Alt+a), /Command *arrow*>Exit Application, /ok, & close switchboards & switchboard manager
Under Objects / Forms, Right / Switchboard>Rename & type frmSwitchboard (so you know that this switchboard is for your forms). //frmSwitchboard & view & test it. You can create a 2nd switchboard aside from your main, default, if you need more. Perhaps you need another switchboard for your reports… Tools>Database Utilities>Switchboard Manager, /New & type **Reports Switchboard**, /ok. If you have 2 reports you'd like to add to this switchboard: select 'Reports Switchb' & /Edit (b),
1. /New & type the **Name of &Report**, (shortcut key, Alt+r), /Command *arrow*>Open Report, /Report *arrow*>choose your report, /ok
2. /New & type the **Name of &bReport**, (shortcut key, Alt+b), /Command *arrow*>Open Report, /Report *arrow*>choose your report, /ok
3. /New & type **Back to &Main**, (shortcut key, Alt+m), /Command *arrow*>Go to Switchb, /Switchb *arrow*>Main Switchboard, /Close

<u>Modifying</u>: to change the label of your main switchb: Open rptSwitchboard in Design View, >Right / defaultLabel>Properties & change the Caption to **new name** ® & you also need to rename the Label's Shadow to your 'new name' too: Notice the 1st option on Formatting *toolbar* (called the "Object List") that \ currently displays "Label 1," /its *arrow*>Label 2 & in properties find & change its Caption to **new name** ®

<u>Picture</u>: to insert a picture: On Formatting *toolbar* /Object List *arrow*>Picture & in Properties: /All *tab*, /in Picture *field* & /its Build (b)>browse & //your picture

<u>Property Sheet</u>: you can prevent this from opening by bringing up the Form's Properties: Right / dark gray area of your form>Properties, /All *tab* & scroll down & //Allow Design Changes>Design View Only

<u>Startup</u>: to have your main switchboard open automatically on user opening the database: Tools> Startup, & type **Name of your business**, /Application Icon's Build (b)>browse & //any icon .ico you'd like to add to your switchboard's title bar, /Display Form Page *arrow*>Switchboard, & uncheck 'Display Database Window,' /ok. To test close & reopen your database... Press F11 to open your database

<u>Modify</u>: to make sure when you close your database with the switchboard any hidden windows i.e. >the database & adding other low security features: Tools>Startup & uncheck all 5 boxes at bottom of Startup *window*, Right / Switchboard. frm>Design View, Bring up Form Properties, /Format *tab* & //Close Button>No (this will disable the switchboard's Red X close (b) when you reopen database) Test your modifications by closing & reopening your database & note:

1. user can't / Red X close (b) on switchboard
2. F11 key has been disabled
3. Right / on Switchboard>Design, disabled
4. Tools>Startup, disabled

Close the Database & this time hold SHIFT *key* & //your database to reopen to disable the disabled items

Data Access Page – to create a web page: Under Objects, /Pages & //Create Page Data Access Page in Design, in Field List find the name of a table you'd like to base your Web page on & / its plus + sign to expand it & drag the fields from the list to the page grid, /ok. /Save & name your web page (This saves a web page in <u>Internet Explorer.htm</u> & creates a shortcut *icon* under 'Objects' in Pages)

<u>Dropdown List</u>: Scenario: let's say you have an Employee ID # field in one table & the names of your employees in another table. To display both the EmpID field & their names from separate tables in a combo box, first create a query that has EmpID field & another that concatenates (combines) the first & last name field. Now create a Dropdown List that will list concatenate names In field list expand a table to reveal the field EmpID

1. On Toolbox / Dropdown List (b) & from field list / & drag EmpID onto Access Page (this creates the EmpID dropdown field, but without any functionality, you'll have to add that next...)

Mouse click /	To >	Task Pane <u>TP</u>:
Button (b)	Enter or Return®	

2. Right / field you just added to your page>Element Properties, /Data *tab* & /Control Source *arrow*>EmpID, /List Row Source *arrow*>your concatenate query, /List Display Field>field from the query that concateneates EmpInfo & /List Bound Field>field from the query EmpID (this binds the EmpID from the table to the EmpID in the concatenated query) & Width>Auto (this will do an auto fit to the largest line of text), /View (b)...

Dropdown List 2: Scenario 2: let's save you have in your customer table a field that has your customers listed as seasonal clients A, W, S, Su for Autumn, Winter, Spring & Summer. To display those codes in a Dropdown List...

   1. On Toolbox /Drop-down list (b) & /in Access Page: & select 'I will type in the values I want,' /Next (Note: most # of columns allowed is 2) in 1st row *cell* type **A** (for Autumn), 2nd row *cell* **W** (Winter), 3rd row *cell* **S** (Spring), 4th row *cell* **Su** (Summer), /Next & for label type **Seasonal**, /Finish

   2. Right / Seasonal Text box>Element Properties, /Control Source *arrow*>the name of the field in you the table containing your seasonal codes, in Default Value *field* type **Su**, /Width *field's arrow*>Auto (Note: List Row, List Display & List Bound Field Properties are blank)...

Browser: to view Access Pages in Browser to preview: File>Web Page Preview

PivotTable: to create a Pivot Table in Data Access Page tied together related fields:

1. Open Field *list* & expand drag a few fields from one table onto your Access Page.

2. In Field *list* expand Related tables *folder* & from that choose a table you'd like as a Pivot Table & / & drag it on to Page & select PivotTable, /ok

Themes: to add a preformatted design to your web page: Format>Theme & select one, /ok.

Hide Records: to prevent user from viewing all records already entered in database, except those the user enters in themselves: Edit>Select Page & in Property Sheet, /Data tab & /Data Entry *arrow*>True

Split Database – to split a database so that you can put your tables on the backend server & the other Objects: Forms, Reports etc. (frontend) are located on the user's machines.

   1. Pro: The admin has only one copy of the database to manage & protect. Doesn't have to visit each users machine to keep them updated.

   2. Con: Database must be set so there won't be bottle necking & lock the database up if a large # of users hit database at same time, & users need read/write access to Tables so thorough security measures must be set.

Open your database & Tools>Database Utilities>Database Splitter, /Split, (back end Default Name is same name but has added '_be.mdb' to name), /ok (Split may take time depending on size of database & speed of computer).

   1. (Note: Tables with arrows to left of them mean they are linked to backend)

   2. (Note: Create your Switchboards AFTER you Split because the switchboard table has to reside on frontend & not with the rest of the tables on the backend)

Security – to protect both backend & frontend after a database has been split

   1. Encryption Decryption – Compacts your database & renders it unreadable by a word processor or utility program (good for storage or electronic transfer)

   2. Hide Objects – to hide database's object, but easy for anybody to unhide

   3. Startup – to specify what user sees when database is opened

4. <u>Password</u> – to set an encrypted password for ONLY opening the database (Note: this can't be used if you ever plan to replicate the database)

5. <u>VBA Code</u> – to keep unauthorized users from editing, cutting, pasting, copying, exporting, or deleting your VBA code.

6. <u>Data Access Page</u> – to set Internet Explorer's settings to prevent unauthorized access to data access page

<u>Frontend Security</u>: When you install access you're automatically made a member of a group named "admins" with a user account named "admin." By default, you also have an empty password & Personal Identifier. Until you activate the logon procedure, the Admin user account remains hidden: Open your Frontend database (database you used to split), Tools>Security>User-Level Security Wizard (Note: the wizard creates an unsecured backup of the database, because if you forget your password you're TOAST! And you will have to recreate your database over again), /Next, /Next & let them all be checked so ALL are secured, /Next & select each Group & read about it (The Admin Group is included by default) & check boxes accordingly, /Next & select accordingly, /Next & select 'Add New User' & in User Name type **Person's Name** & a password for them (Note: password is case Sensitive & be sure to write passwords down!), /Add this User to the List (b), /Next & add your name to the 'Admin' *group*, then add the other users to your choice of Users *groups*, /Next, /Finish & read the report... (Note: the report has user names & their passwords too! Saved as a .snp or Snapshot file. Store it in a save place) Close the Snapshot file & /ok to accept encryption & close Access. On Desktop find your database's shortcut & //it (Note: database is secured, but your backend database database_be.mdb is an unsecured copy of your database.mdb) & log on as the admin. File>close (don't exit out of Access or you'll be exiting out as admin too & if you do, you won't be able to do Backend Security)...

<u>Backend Security</u>: File>open database_be.mdb & Tools>Security>User-Level Security Wizard, make sure 'Modify...' is selected (If not that means you've go to //Shortcut to database.mdb (frontend) on desktop & log on as admin & FILE>CLOSE & FILE>OPEN the database_be.mdb (backend) or these next steps won't work), /Next & make sure all Objects are selected, /Next & check: New Data Users, Read-Only Users & Update Users, /Next, /Next, /Next, /Finish File>X & /Yes to save as Snapshot file & close Snapshot. Tools>Security>User and Group Permissions, selected Radio option 'Groups' & select 'a group' & assign or remove permissions, /ok

Convert 2000 to XP – to convert Access 2000 database> XP: Tools>Database Utilities>Convert Database>Access XP, select database.mdb, /Convert & name it, /Save, /ok.

Protect VBA Code – to prevent unauthorized users from deleting, editing, copying your VBA code: Select any form in Database & on *toolbar* /Code (b) & Tools>VBA Example Properties, /Protection *tab* & /Lock & for password type password, /ok.

Password – to add a password to database so when emailed, if intercepted, no worries. Also, this the 1st line of security defense and if they break through that you'll have workgroups, switchboards &

Mouse click / Button (b)	To > Enter or Return®	Task Pane <u>TP</u>:

permissions: Open Access 1ˢᵗ & log on as admin & then file>X the database & then file>open & highlight yourdatabase.mdb (you must log on as admin first & close the database without exiting the Access program so Access can retain admin privileges & by-the-way is a warning to when you're an Admin ALWAYS exit Access completely or somebody else can log on as you), /Open *arrow*>Open Exclusive & Tools>Security>Set Database Password & type in password <u>Remove</u>: File>Open, highlight database.mdb, /Open *arrow*>Exclusive & Tools>Security> Unset Database Password.

Encryption – Encryption compacts the database & makes it great for storing or sending your database electronically (it's also good as a 2ⁿᵈ level of security aside from having a password): With only Access opened (not databases): Tools> Security>Encrypt Decrypt Database, select database.mdb, /ok, select database.mdb again & /Save, /Yes.
<u>Decrypt</u>: to Decrypt follow same procedure as above…

MDE Extension – turning .mdb files into .mde files to protect your database from someone copying the application & the supporting codes. .mde files prevent users from:
1. Viewing, modifying or creating <u>forms</u> in Design View
2. Viewing, modifying or creating <u>reports</u> in Design View
3. Viewing, modifying or creating <u>modules</u> in Design View
4. Add, delete or change references to database or object libraries
5. Changing source code
6. Import or export forms, reports or code modules

Qualifying factors in order to save your .mdb as .mde file:
1. Must remove replication if any to the database
2. Have complete Admin privileges

Access Program (no databases open): Tools>Database Utilites>Make MDE File & select database.mdb, /Made MDE (b), /Save.

Distribute Database – to backup & then distribute Frontend (to users) & Backend (to server) databases after they've have security measures applied to them from steps above. Note these following steps are for PC networking without a server, but same concepts are easier applied to a server:
1. Copy database_be.mdb (including its shortcut) & Secured.mdw (mdw file stores info about the members of the group) to a shared directory *folder*
2. <u>On User's PCs</u>: Open *any* folder & Tools>Map Network Drive & pick a letter i.e. Z as the drive letter of the shared directory folder.
3. <u>On Administrator's PC</u>: On Desktop //database.mdb's *shortcut* (frontend) & log on as Administrator & Tools>Database Utilities>Linked Table Manager (to re-link this master copy to the mapped drive Z:) & check all tables & check 'Always Prompt for New Location,' /ok & browse to mapped drive 'Z' & //database_be.mdb (backend)
    i. Copy the re-linked App.mdb AND App.mdb's *shortcut* into the shared directory & then from the shared directory copy the App.mdb AND App. mdb's *shortcut* onto each user's PC
    ii. Remap the database.mdb *shortcut's* Target *address* 1ˢᵗ to User's own computer's Access Database folder MSACCESS.EXE, 2ⁿᵈ to the own user's desktop storing a copy of the frontend database, 3ʳᵈ to the Secured.

mdw file in same shared directory drive. For example, if we created a Z drive on a User's computer the shortcut's…

1. Target to the database.mdb file would be as follows (In Windows XP of course): **"C:\Program Files\Microsoft Office\Office10\ MSACCESS.EXE" "C:\Doccuments And Settings\User Name\ Desktop\App.mdb" /WRKGRP "Z:\Secured.mdw"** AND the
2. Start In **Z:**

Link Table – inserting a table into a database from another: Open database1.mdb & File>Get External Data>Link Tables, //database2.mdb, & select a table, /ok.

Link to Excel – to link Access to Excel: Right / in White database area>Link Tables, /type *arrow*> Microsoft Excel, //Excel Workbook.xls, & select a sheet & /Next, check 1st row contains column headings, /Next & type a name, /Finish, /ok.

Update Links – to check that your tables are linked to correct source: Tools>Database Utilities>Linked Table Manager, /Select All (b), check 'Always prompt for new location,' /ok, //database.mdb, /ok, /Close.

Refresh Links – to update your table to backend source: Right / any table>Linked Table Manager, /Select All, /ok.

Import XML Data & Structure – is Extensible Markup Language and is used to describe & deliver data on web. It separates the data layer from the presentation layer so data can be presented in multiple ways & hence leaves the interpretation of data up to the application that reads it. It's becoming a more popular platform to share info between different applications. To import XML: Right / white database area> Import, /data type *arrow*>xml, //Name.xml, /Options & select 'Structure & Data,' /ok, /ok… //your Name *table*.

Export XML – to export so others can open & read data without loss because their platform is Independent: Select a table & Right / it>Export, /type *arrow*>xml, /Export (b)…
- XML – only data, no info on how data is to be presented
- XSD – writes Schemas for dataset & contains info about structure of data
- XSL – data presentation…

/Advanced (b), /Schema *tab* & uncheck 'Include Primary Key,' /ok
Export Report as HTML: Right / a report>Export, /type *arrow*>XML, /Export, /ok.

Replicate – the concept here being able to create a copy of database, make changes on the road, come back & update the Master from it (Note: the replica of the original can only have records updated, not designs. While in original you can make changes on ALL!: Open database.mdb & Tools>Replication> Create Replica, /Yes, /Yes & name it, /ok, /ok (Note: yellow *icons*, & you're not in the Replica, but back in the Design Master. Close the Design Master, copy

Mouse click /	To >	Task Pane TP:
Button (b)	Enter or Return®	

the Replica of the Master to your laptop & make all the changes you want with the records. When you're back at the office you can synchronize your replica with original & ANY changes made on either side of the Master or Replica will update in the other i.e. **Open the Replica database.mdb, Tools>Replication>Synchronize Now, browse to find the Master database.mdb & //it, /ok, /ok**

Extras

Store Pictures: How to store a different picture for each record

1. In your table create a *field* title 'Picture' & / its Data Type arrow>OLE Object
2. /View (b) (to go to Datasheet View) & / in the Picture's *field* or *cell* of the record, & Insert>Object & browse to find & insert your picture. Now go to record two's Picture *field* & Insert>Object & browse to find & insert your picture and continue this for each record you'd like a picture for (Note: There will be no actual pictures in these fields, they will come later when you view these fields in a Form).
3. Close Table
4. Open Form -
   a. Create a Form based upon your table that includes your Picture *field* & then…
   b. On Toolbox /Image (b) and then in your form's field list, / & drag the field you created in your table for the pictures, on to your Form, and you're done!

Copy Field Macro: to copy a field in one form & paste in a field in another Say you have Customer Form that contains a subform – containing several address records for each client. A client calls to have their lawn mowed at one of their many addresses. I want to //on any address in subform & have that address copied & pasted into an already created street field (in Orders Table) in Orders Form using a macro: In Design View of Customer Form, Right / Address *field*>Properties, /Event *tab*, /On Double Click field's Build \ (b)>Macro Builder, Name Macro **mcrCopy to Form** & add the following actions with their properties:

1. GoToControl
   a. [any field but the main one you want to copy] (it's ironic that access won't go to the first field & later copy unless you set up 2 gotocontrols & copy on the 2ⁿᵈ field ????? beats me…
2. GoToControl
   a. [Addresses]
3. RunCommand
   a. Copy
4. OpenForm
   a. Form Name = Orders
   b. View = Form
5. GoToRecord
   a. [Street] (make sure you have this field in your Orders Form (based upon your Orders Table) that will store the addresses copied from.
6. RunCommand
   a. Paste
7. RunCommand
   a. SaveRecord

Shortcuts

F11 – Brings database window to front
Ctrl+F11 – Toggles between custom menu & built-in menu
Ctrl+G – Brings up Immediate window
Ctrl+Break – Pause the application
Alt+11 – Opens Visual Basic

Mouse click / Button (b)	To > Enter or Return®	Task Pane TP:

Excel

Editing Cell – // the cell you want to edit its text, or / in cell & then / up in Formula *bar*.

Copy or Move Info – To move or copy data from 1 cell to another:
- Select the cell & on *toolbar* /Cut (b) or /Copy (b) & then / in a different cell & on *toolbar* /Paste (b) (Press ESC *key* to get rid of marching ants) OR,
- / the cell's border & drag; also if you Cntrl+/ & drag the cell's border it will make a duplicate of that cell.

AutoFill – Or the black cross +, used to repeat a cell or a range of cells' patterns: Move mouse over lower- right corner of cell or a range of cells & drag black cross + down or right depending on which direction you prefer your info patterned.

Text to Speech – Excel can read back your cells' data to you: 1st select a range of cells you want read back to you, then in *menu* View>Toolbars>Text To Speech & on the *toolbar* press Play (b).

Formulas – always begins with = i.e. =d3+d4+d5
   =max(A5:B6) (this formula finds the highest # in range A5 through B6)
   =min(A5:B6) (this finds the lowest # in the range)
   =count(A5:B6) (this counts up the # of cell in the range)
   =sum(A5:B6) (this adds us the range)

Absolute References – Prevents a reference from changing when you copy & paste the formula of 1 cell into another i.e. if you have one cell that will be divided by all the rest, that 'one cell' is a constant & you want that once cell's reference to *stay!* Before you copy or AutoFill the formula to another cell, up in formula bar, place cursor before cell reference you want to *stay* & press F4 (it will add $ or Absolute Values to your formula. Easy way to remember this is Fido! When you want to tell your dog Fido to stay or in this case a cell called an 'Absolute Reference' there are **4** letters in Fido & it begins with the letter **F, F4!**)

Formatting Cells – Changing the cells color, font, # style, border (can outline a cell in color), or alignment: Right /cell>Format cells.
   Custom Format: Right / cell>Formatting cells, /Number *tab* & select Custom, / in Type *field* & use quote marks "" around the text you want as a part of your format i.e. "SS" for Social Security.
   Clear Formats: Edit>Clear>All
   Style: To save a format style. Select the cell with the style in it you want to save &: *Menu* Format> Style & type in Style Name *box* a name, /ok. To apply style, Format>Style & /Style Name box *arrow*>any style, /ok.
   Cell Borders: To outline a cell with a black border. *Toolbar*, /on Outside Border (b).

Find & Replace – To find & or replace text or numbers in a cell: Edit>Find & type in what it is you want to find & /Next. If you need to find certain formatted cell or text with certain styles Edit>Find & type in your #s or text & then /Options (b) & /Format (b) & select a particular format to those #s or text you're trying to find & /Find. For find & replace type in what you're finding with its text & /Replace *tab* & type in what you want it replaced with including any special Format (Warning! Once set the Format (b) to find a certain format the Format will not clear after you find it; so if you want to find something else /the arrow to right of Format's (b)>Clear Format).

Merge Cells – Select range & up in *toolbar* /Merge & Center (b) (To unmerge cells: select merge cells & /Merge & Center (b) again).

Worksheets – To manipulate the worksheets…
    <u>Sheet Tabs:</u> Delete, Insert, color, move, copy or rename worksheets: Right / worksheet *tab*.
    <u>Moving:</u> To change orders of worksheets: / & drag the sheet tab before or after another tab.
    <u>Copy:</u> To duplicate a worksheet: Hold Ctrl, / & drag that sheet's *tab*.

Charts – To create a chart: 1st select your data & on *toolbar* /Chart Wizard (b) & follow Wizard's instructions.
    <u>Chart Options:</u> Add a legend, labels or axis titles: *Menu* Chart>Chart Options. (Note: Make sure your chart is selected otherwise there will be no "Chart" *menu*)
    <u>Chart Type:</u> Change current chart into another i.e. column to pie: Select Chart & Chart>Chart Type.

Freeze Cells – You can freeze rows or columns or both at same time…
    <u>Rows:</u> To keep a fixed set of rows frozen when you scroll down your spreadsheet: / above the row you want frozen but must be in A *column*, then in *menu* Window>Freeze Panes…
        <u>Unfreeze:</u> Window>Unfreeze (Note: whatever cell you select everything will freeze above & to the left of it).
    <u>Columns:</u> To keep a fixed set of columns frozen when you scroll right in your spreadsheet: after the column you want frozen but keep it in 1 *row*, then in *menu* Window>Freeze Panes (Note: whatever cell you select everything will freeze above & to the left of it).

Repeating Rows – To have a select set of rows repeated & printed on each printable sheet: File>Page Setup, /Sheet *tab* & /Collapse Dialogue (b) for "Rows to repeat at top" & select your Rows (Note: rows can only be selected in together, not sporadically here & there in this feature), /ok.

Headers Footers – To put duplicate text in the headings or footings of each printable sheet: View>Header and Footer & /Customer Header or Footer (b) & type in what you'd like in 1 of 3 *sections*.

Mouse click /	To >	Task Pane <u>TP:</u>
Button (b)	Enter or Return®	

Page Break – If your sheet isn't fitting correctly in your print preview: In Print Preview, /Page Break Preview on *toolbar*, then find & / & drag dark, dashed blue line on the page to where you want page to break. When finished, to get back to normal view, *menu* View>Normal View.

Print Range – Prints an area specified. Select your area you want printed 1st, then File>Print Area> Set Print Area (To Clear Print Area: File>Print>Clear Print Area).

Inserting Columns & Rows – Right / on a row (1, 2, 3 etc.) or column (A, B, C etc.) *header*>Insert or delete.

Inserting Cells – Select a range of cells & Right / them>Insert (Shifts cells selected down or right) or delete (deletes selection & them pulls other cells into its deleted range up or left).

Quick Calculations – To get a quick summary or average use the Status bar. 1st Select a range of cells, then down at the bottom Right / on Status bar>Sum or whatever it is you want a quick calculation of.

AutoSum – For quick sum of cells. / in a blank cell adjacent to the range of cells you want to add & on *toolbar* & /AutoSum (b) on *toolbar*, make sure the marching ants select your range you want summed, if not use mouse to / & drag your new select of cells & hit ®.

Import from Web – To import tables into Excel from Internet: Data>Import External Data>New Web Query & type in desired web site address. /A *yellow tag* to select a table, /Import (b), /any cell to import your data, /ok
Refresh: To refresh imported data from web: 1st /in cell were you imported your data & on External Data *toolbar* /Refresh (b).

Set Print Area – Select your range of Data & File>Print Area>Set Print Area, /Print Preview (b)… To Clear: File>Print Area>Clear.

Import Access – Data>Get External Data>New Database Query, & select MS Access Database ® & select your database ® From Available *list* select a table & /the right add arrow (b), /Next, /Next, /Finish… /ok.

Hide Data – To hide rows or columns from printing or exporting to another application program: select column or row *headers* & right / it or them>Hide. To unhide select the column before and after the hidden column & right them both to unhide: i.e. if column B was hidden then select column A & C and right /them both>Unhide.

Export to Word – File>Save as & /File Type *arrow*>Tab Delimited, /Save, /ok, /Yes & File>X, /No. Open Word & File>Open, /File Type *arrow*>All, //Export.txt.

Interactive Web Page – Select a range of cells you'd like to have saved in a web page & Files> Save as Web Page: select 'Selection,' /Add Interactivity, /Change Title (b) & type in title name of your web page, /ok & change File Name any name you want, /Publish, /Open Published Web Page in Browser, /Publish.

Create Name Ranges – To give a cell or a range of cells a name: Select a cell or a range & up in Name Box (left of Formula bar) /in it and type in a new name & hit® Now you can /in a blank area and then /Name Box's *arrow*>your name and it will take you to it. To delete a Named cell or range: Insert>Name>Define & select your name & /Delete.

Link Data – Pulling data from one worksheet and linking it in another; so when you change data in one worksheet it updates in the other(s): /in a blank cell and type = then /on worksheet B and select a cell that has some data & hit® & any change you make in that cell on worksheet B will up date in worksheet A. Formulas work same way i.e. in worksheet A /in a blank cell and type **=sum(**Then / on worksheet B and using the Ctrl *key* select multiple cells you'd like to add to your SUM *formula* & when done hit® Again, any changes you make on worksheet B that is linked (using the equals sign links them), will update in your formula on worksheet A.

IF Function – Calculated test & if true you can program it to display one thing or if false programmed to display something else. Syntax for function is IF(logical_test, value_if_true, value_if_false): /in a blank cell & on *toolbar* /fx (b) & /Or select a category *arrow*>Logical & select "IF", /ok. In top *field* type in a 'Logical test' i.e. A6>33 (the test is whatever is in cell A6 has to be greater than 33), press tab key & type what you want if its true A6*22% (for true, you said to multiply whatever is in cell A6 by 22%), press tab key & type in 0 or in quotes "You are soo Fired!" (for false, that cell A6 is not greater than 33 then either 0 bonus or they get "Fired!").

PMT – To calculate payments on a loan: To set up assign 3 cells to contain Interest payment (say A5), another cell assigned for # of payments to be made (say A6), and assign the total loan (say A7). / in a cell to figure the 3 variables (say A8) & /f(x) (b), /Or select a category *arrow*> Financial & select PMT *function*, /ok. In first field enter interest rate **A5/12** (A5 is Divided by 12 months), 2nd field enter # of payments **A6**, 3rd field enter total loan **–A7**, /ok. (Note: now you can change cells A5-A7 with different #'s to come out with different results in A8; your monthly payments)

Auditing Toolbar – Used to trace values to a formula or vice versa: Tools>Auditing>Show Auditing *toolbar*. /in a cell that a formula depends or is based upon & on *toolbar* /*Dependents*. /in a cell containing a formula & on *toolbar* /*Precedents*. To clear / Remove All Arrows (b).

VLookup – Vertical Lookup; to lookup up a value in a table and return its compliment value. Syntax: vlookup (<u>what</u> to lookup, <u>where</u> to look it up like a table (looking in left most column & checking

Mouse click /	To >	Task Pane <u>TP</u>:
Button (b)	Enter or Return®	

if it's greater than the *what*), <u>which</u> columns contain desired info.(if *what is* greater than the values in *where* you're looking up then it will return the value in same row from a column you specified): Create a two column table with many rows, and in its 1ˢᵗ row (the dead row) type in your column headings & below the dead row in the left column type your totals you'd like later compared & in right column type in the values you'd like returned from your lookup. Select all the cells in your table except the dead row & in up *toolbar* in Namebox type a name for your table & hit enter.

- /in a cell you that has a number you want compared to in your new table & on *toolbar* /fx (b) & /Or select a category *arrow*>Lookup & under References select VLOOKUP, /ok…Type the cells reference i.e. A4, you want compared to your table, press *tab* & type in name of your table, press *tab* and type in the column number you want to retrieve the data, (Note: you could use the Last *field* "Range Lookup" to type False, and it will bring back only exact matches; in other words, the lookup has to match the number in the left most column of the table otherwise it displays N/A), /ok..

Pivot Table & Chart – Performs like AutoFilter, sets up your database to filter through it: /in the middle of your database & *Menu* Data>PivotTable & PivotChart Report, /Next, /Next, /Finish. Next / & drag your column *headers* onto their appropriate fields (This is guess work as to how you want your Pivot Table to operate and filter. If you put a field in the wrong place, / on that field that's in the wrong cell & drag it out of there and let go of the mouse and try again with some other field).
PivotChart: /Chart Wizard (b) to create a PivotChart
Refresh: /on your Data Sheet & change your data, then go back to your Pivot Table & on PivotTable *toolbar*, /Refresh (b).

Now – A function used to always enter in today's date every time you open Excel: / in a cell you always want to show today's day & type **=now()** ® (Enter's current date & time).

Scenario – To create your own scenarios and view them; to toggle quickly through the views too:
1. First create a scenario view (i.e. marketing or budget view), and then select all and any ranges that contain values that are a part of your scenarios projections. *Menu* Tools>Scenarios, /Add, Name *box* type a name for your base or original scenario and in Comments *box* type any necessary comments, /ok, /ok, /Close.
2. Next *Menu* Tools>Scenarios, /Add, & in Name *box* type the name of your new scenario, /ok, then scroll and find your cells that you'd like to make your new projections /changes to, /ok, /Close
   a. View Scenarios: *Menu* Tools>Customize, /Commands *tab*, under "Categories" select "Tools", & in right pane find, / and drag "Scenario drop-down list" to left of "Undo" on *toolbar*, /Close & select from drop-down list each scenario.

Solver – You know the answer you want, but not the inputs; solver can help when you have more than one variable input. Solver is based on PMT *function* found earlier in these notes: Tools>Add Ins & check "Solver Add-in" If you wanted to buy a truck with a monthly payment of $555 and a max term of 52 months, solver can help: (solver can't be perfect, but can come close to desired amounts). /in the cell that contains your PMT *formula* result & Tools>Solver, select "Value of"& in its *field* type our desired monthly payment of 555, now we need to select the variables or cells solver can change: /By Changing Cells *collapse dialog* (b) & /in the cell containing your Principle Amount type **,** & /in a cell containing your # of payments ®, /Add, /in the same cell containing # of payments & type in "Constraint" *field* 52 (as we don't want the term of our loan to be more than 52 months), /ok, /Solve & if we like the results /ok.

Goal Seek – Like Solver, but it solves problems with only 1 variable & is based on PMT *function* too i.e. to solve a formula based on the value that you want the formula to return: For example: let's say the only variable we want in our PMT *formula* will be the # of payments made in terms of months because we don't care about how many months we'll be paying for our truck because we have to have it! /in cell containing your PMT *formula* & Tools>Goal Seek, in 'To Value' enter a monthly payment, t, /By changing cell's collapse dialogue box (b), /in cell containing your # of payments ®, /ok.

Database – To sort effectively & correctly there needs to be a distinction in your database you created so Excel can tell what to sort or not. If done incorrectly the column sorted won't keep its sister information in adjacent column(s) (same row), glued when sorted…

Design Flaws: Things to *avoid* when creating a database so Excel can distinguish that you have a database when sorting it or other actions:

1. Different Formatting within a column (all data must be same format)
2. Mixing "text" & "values" in same column (must be one or other)
3. No blank row (take your chances)
4. Border separation (use formatting for borders, not Drawing *toolbar*)
5. Column Label formatting isn't different from rest of data
6. Don't add spaces in cells (use alignment (b)s)

Sorting: To sort data ascending (A to Z) or descending (Z to A): /in *column* you want sorted & up on *toolbar* /A – Z or Z – A.

Multilevel Sorting: To sort more than one *column*: *Menu* Data>Sort

Consecutive Sorts: Allows you to sort your lists by more than 3 columns. Sort the least important *columns* 1st & then sort again. To Sort 5 *columns*: Data>Sort & use all 3 Sort fields & then repeat; Data>Sort & use 2 Sort fields.

Months – Months by default are sorted by Excel as Apr, Aug, Dec etc., but to sort months as Jan, Feb, Mar etc.: /inside Month *column, Menu* Data>Sort: /Options (b), /First key sort order *arrow>* January, February, March, April (Note: if your months are abbreviated as Jan, Feb, Mar. etc. then /First key sort order *arrow>*Jan, Feb, Mar. otherwise it won't work, formatting must be kept same).

Subtotals – To subtotal a column that has been previously sorted ascending (A to Z). If there is any duplicate data in that column (or groups), then you can subtotal those groups: *Menu*

Mouse click / Button (b)	To > Enter or Return®	Task Pane TP:

Data>Subtotals, /At each change in *arrow*>"the name of your column" you have duplicates or groups in, /ok. To Remove: Data>Subtotals, /Remove All.

AutoFilter – To have column headers display drop-down arrows that can be used to filter the database: *Menu* Data>Filter>AutoFilter. To clear Filtered *columns* to re-filter: Data>Filter>Show All
To remove AutoFilter: *Menu* Data>Filter>AutoFilter
Wild Card: To use * to help filter a column i.e. if you want all last names that begin with the letter D*: / the column's *header* arrow>Custom & /in the box to the right of 'Equals' & type in a wild card D*, /ok (asterisks means anything, hence show all name that begin with letter D, but doesn't matter what comes after it).

Hyperlinks – Right/any cell you want linked to another application or workbook>Hyperlink, //a file. To edit Links Name: /in another cell, now to edit the name of the hyperlink / & hold the mouse on cell with hyperlink until you see Hand turn>White Cross, then you've selected the cell so you can now change Employee name to **Update Loans** in Formula *bar*.

Comments – Adding comments to a cell when that pop up when you hover your mouse over them: Right / any cell>Insert Comment & type in your comment & when finished /in a blank cell. To Edit comments: Right/cell will comment>Edit. To Delete comments: Right/ cell>Delete.

Template – To save your creation into an original that can't be touched, but copied: File>Save As & /Save as type *arrow*>Template (*.xlt), type in a file name & /Save. To get a copy of template: File>New & //your template.

Conditional Formatting – to apply formatting to cells when criteria is met i.e. a range of cells for sales is less than $10,000: Select the range of cells & Format>Conditional Format, /2ⁿᵈ *arrow*>Less Than & in last *field* type **10000**, /Format (b) & select your formatting i.e. color red, /ok, /ok.

Validation – This is a rule that locks a cell(s) to accept only certain #s i.e. say when it came to ordering products like toothpaste the minimum order is 5 & max is 27 tubes:
1. 1ˢᵗ select the cell or range of cells you'd like this rule to apply to,
2. Data>Validation: /Allow *arrow*>Whole # & for minimum type **5**, & maximum type **27**, /Input Message *tab*, /Title *field* & type the name or your error, press tab & type in your error message.
Remove: Select cell(s) & Data>Validation & /Clear all
List: To create a cell that contains a list of #'s for a user to choose from: First enter the #'s you want as a list in their individual cells somewhere in your spreadsheet i.e. #'s like 1, 2, 3 & 4. Then / in a blank cell you'd like those #'s displayed as a list & Data>Validation, /Allow *arrow*>List & /Source collapse dialogue box (b) (or tiny red *arrow*) and using the mouse select the range of your #'s 1, 2, 3 & 4 & hit ®

& /ok. Now / in your blank cell you designated to contain your list & then /it's *arrow*>1, 2, 3 or 4.

Customize Menus – to create your own menu with menu items like 'Save as' & 'Copy': Tools>Customize, /Commands *tab*, under 'Categories' select 'New Menu' & in right *pane* / & drag 'New Menu' up on your Menu *bar*, to the right of 'Help' (b) Under 'Categories' scroll & /File & in right *pane* drag "Save as" onto your New Menu. In 'Categories' /Edit, in right *pane* drag Copy to Menu. To name your menu: Right / your Menu>& change Name to a new name, Close the *panel* & test your new menu…
Remove: Tools>Customize & drag your New menu off into a blank area, Close *panel*.

Custom Toolbar – to create your own customer toolbar with toolbar items like 'Save as' & 'Copy': Tools> Customize, /Toolbars *tab*, /New (b) & name your toolbar, /ok, /Commands *tab*, under 'Categories' /File & in right *pane* / & drag 'Save as' onto your new toolbar. Under 'Categories' /Edit, in right *pane* / & drag 'Copy' onto toolbar. Close *panel* & test.
Remove: to remove toolbar: Tools>Customize, /Toolbars *tab* & either uncheck your toolbar or select it & /Delete (b), Close *panel*. To remove Buttons: Tools>Customize, & on any toolbar you can / & drag their buttons off in to a blank area & let go.
Reset: If you accidentally remove a (b) from the standard or formatting *toolbars* you can bring them all back when you Reset them: Tools>Customize, /Toolbar *tab*, select the toolbar you're missing buttons from either Standard or Formatting & /Reset (b), Close *panel*.

Macro – A simple way to record all your repetitive steps in to a simple macro that when ran executes all of your steps like lightning: 1st start recording your steps *Menu* Tools>Macro>Record New Macro, type in a name for your macro (you can add it as a shortcut *key* if you type in a letter like q, but if you hold down shift & type q your shortcut for that macro will be Ctrl+Shift+q), /ok; now go through all your step & don't worry if a mistake is made and stop recording, but correct it and keep going i.e. Right / a cell>Format Cell, /Pattern *tab* & select a light color, /ok; on the little floating *toolbar* /Stop.
Run Macro: To apply all your recorded steps onto a cell or range: /in a cell you want your previously recorded macro color applied too *Menu* Tools>Macro>Macros, select your Macro & /Run.
Delete Macro: Unless you know Visual Basic programming never try to edit your Macro, but delete it and re-record it: *Menu* Tools>Macro>Macros, select your Macro & / Delete.

Mouse click / Button (b)	To > Enter or Return®	Task Pane TP:

Outline – is a structure you add to your data, up to 8 levels. You can show or hide each level of detail to help control viewing large amounts of data. Make sure your data has an easy to follow outline that will be easy for Excel to pattern and form groups: Data> Data>Group & Outline>Auto Outline. To clear, Data> Data>Group & Outline>Clear Outline.

Workspace – if you need to open more than one workbook at a time you can save them as a workspace: Open up 2 or more workbooks & File>Save Workspace as name it, /Save. Close all workbooks & file>open your workspace file.xlw.

Consolidating – Consolidating data from more than one worksheet or book. In other words, taking values from multiple cells & cramming it into once cell that can be expanded to reveal the values from other cells contributing to the total or consolidated cell; or left collapsed to simply display the total:

Workbook: Consolidating data from multiple books: 1st all your books must be open that you want consolidated, then go to your master book's sheet that you want your results displayed & /in a cell, Data>Consolidate, /Browse, & find your 1 of many books and //it, in reference *box* place cursor right at the end of that file name & type the address for your consolidated link (you already have the book) that includes the worksheet & range of cells to be consolidated i.e. like **Sheet1!A6:B22** (Note: to simplify you can name the range "Spiffy Sales" for the range A6:B22 if it helps before you start Consolidating), /Add… Place cursor back in Reference *box* and replace only the name of your book with a name of another book you want consolidated, /Add & repeat process for as many books needed; then put a check mark in "Create Links To Source Data" (By checking this, it puts a + signs after you /ok into your margins to /on to view the details for those consolidations), /ok

Worksheet: Consolidating data from multiple sheets: /in a cell, Data>Consolidate, /Collapse Dialog (red arrow) & /on another sheet tab you want to consolidate & select the range i.e. A9:B12 & hit enter, /Add, /another *sheet* & select range, /Add, and repeat till your finished & under "Use Labels" check "Left column" (only check this if you have labels to the left of all your data you've selected in your worksheets and want to add those labels next to your consolidated data results), /ok.

Workbooks Links – Linking books in a formula is like linking sheets, but you need to have all the books opened you want linked. In other words, you can link a cell to another cell in another workbook by itself or as part of a sum formula: /in a blank cell & type **=sum(** Window>your other work book you want as a part of your linked sum formula, then /on its sheet's tab & /in the cell that you want to add, then press + & again, Window>your other work book you want as a part of your linked sum formula, then /on its sheet's tab & /in the cell that you want to add & hit ® when finished.

Link Change: To change a link from one book to another (Note: Make sure this new book has the info in the exact same cells you're linking to from the old book): Open the book that you want linked to another aside from its current link, Edit>Links & select the old link & /Change Source, browse & //on your new book you want linked, /ok.

Exporting XML – XML doesn't add tags like when you File>Save as Web Page, because Microsoft would add extra formatting tags proprietary to only FrontPage; xml is clean that can be easily used by other programs including other applications not made by same manufacturer: File>Save as & name your page, /File Type *arrow*>XML Spreadsheet, /Save & File>X.

Import XML – Scenario: web page has data you want in Excel, don't copy & paste because cause of error can be great: In New Blank Workbook: Data>Import External Data>Import Data, //xml workbook.xml, /ok (Note: no formatting imported).

Web Query XML – If you're constantly importing XML to keep up to date you can create a "Query" to that XML file to update automatically daily i.e. say you're company has an xml file that put out on the company's Z: drive for anybody to import to their own computers: Open your workbook & Data>Import External Data>New Web Query & in Address *bar* type **z:/name of folder containing xml data/name of xml file.xml**, /Go & select a table you'd like imported, on *toolbar* /Options & select 'Full HTML,' /ok, /Import, /ok. Refresh: to keep this data updated simply: On External Data *toolbar* /Refresh (b) (& If source has changed, new data would be shown) File>Save as **My Available Commissions** & File>X.

Protect Worksheet – To lock cells from others changing them (By default ALL cells are locked, but are only effective when you *protect* them): Select a range of cells you *want* others to have access to or the ability to change, Right / your selection>Format Cells, /Protection *tab* & remove check from "Locked" & Tools>Protection>Protect Sheet (or workbook). Unprotect: Tools>Protection>Unprotect Sheet.

Protect Workbook – to protect Structure or Worksheets from being edited i.e. to prevent worksheet tabs from being moved: Tools>Protection>Protect Workbook, /ok.

Shared Workbook – to use a network folder to save the spreadsheet on so other can work on it at same time: Open your workbook you'd like to share & Tools>Share Workbook, /Advanced *tab*, for 'Number of days to keep history' enter **10** (keep track of changes), /ok. File>Save as & put it on the Network as with a name, /Save Now other coworkers can open up the network folder & open up your 'Shared Workbook' & start making changes to the specified cells on your shared workbook & then when they're finished: /Save & it will automatically update your workbook with other changes that people are working on at the same time.
Protect Tracking: this protects the change & share history tracking (to have a password, protect with a password, this must be set prior to the step above, Tools>Share Workbook): Tools>Protection> Protect Shared Workbook & check *box*, /ok, /Save
Merging Workbooks: If Bob emailed you a workbook and you'd like to merge his changes in with yours, make sure your Workbook has its sharing turned on as explained above: With your workbook opened, Tools>Compare & Merge Workbooks, & // the workbook to merge
Tracking Changes: To accept or reject merged or shared data: Tools>Track Changes> Accept or Reject Changes, /ok, /Accept (Note: *window* shows Name of

Mouse click / Button (b)	To > Enter or Return®	Task Pane TP:

person who changed it, date/time it changed & what was there before it), view the coworkers #'s & select the ones you want & /Accept.

Chart Non-Adjacent Data – to add data that's in different parts of your spreadsheet to a chart: In your workbook select the ranges using Ctrl *key* i.e. A5:B9 & E5:E9. *Toolbar* /Chart Wizard (b), /Next, /Next, /Title *tab* & in Title enter name of your data, /Next & select 'As New Sheet,' /Finish…

Chart Axis Errors – to show 'potential' % of errors in a "XY (Scatter) *chart*: on XY (Scatter) *chart* Right / a dot>Format Data Series, /X Errors Bars *tab* & select 'Both,' /Patterns *tab*, in Marker *area* /its Style *arrow*>Triangle, /ok

    Location: to change the location of an embedded chart to a Chart Sheet or a Chart Sheet to embedded: >Select an embedded chart & Right / Chart>Location, select "As New Sheet," /ok.

    3D Chart: to add 3D effect: Right / Chart>3D

    Trend Line: a line marking a trend in the chart to predict future trends: Right / chart>Add Trend line, >/Options *tab* & change Forward>**2** (to set trend out to forecast the trend by 2 points), /ok.

Drawing Toolbar – Used to create objects…

    Circles & Squares: to create circles & squares: /on Oval, hold SHIFT *key*, / & drag. To create multiple circles, //Oval… when finished hit ESC *key*.

    Center Objects: To center an object within another object: / on 1st object then hold SHIFT & / on remaining object to be aligned, /Draw>Align or Distribute>Align Center then go back & repeat>Align Middle.

    Send To Back: To send an object on top of another behind it: Select top object, >/Draw>Order>Send to back.

    Selecting Objects: To select objects if original mouse pointer doesn't work: On Drawing *toolbar* /White Arrow & then you can select object, but to go back to selecting text or cells again hit ESC *key*.

    Group: to group objects together i.e. you have two circles you can group them as one: Select the 2 circles using the Shift *key* & on Drawing *toolbar* /Draw>Group

Shortcuts

Ctrl+S – *Save*
Ctrl+A – Selects *all*
Ctrl+Z – Undo any action.
Ctrl+Y – Redo any undid action.
Ctrl+X – *Cut*
Ctrl+C – *Copy*
Ctrl+V – *Paste*
Ctrl+F – *Find*
Ctrl+P – *Print*
Ctrl+/ – Selects cells randomly

<u>Ctrl+Home</u> – Cell A1

<u>Ctrl+End</u> – Goes to end of database, not end of spreadsheet.

<u>F4</u> – Like Format Painter, *but* it only applies to *most recent* format.

<u>F5</u> – To go to a cell or your personalized named cell or range.

<u>F11</u> – To quickly create a Column Chart. After you've selected your data press F11

<u>Page Down & Up</u> *keys* – Pages down/up 24 rows at a time.

<u>Alt+Page Down & Up</u> – Pages to Right/Left 9 columns.

<u>Shift+/</u> – Selects cells in blocks

Mouse click /	To >	Task Pane <u>TP:</u>
Button (b)	Enter or Return®	

Outlook

Send E-mail – In Outlook *bar* /Inbox & *toolbar* /New (b)...

    <u>To</u>: Type in email address in this field or /To (b) & select a name here & /To (b) & /ok

    <u>CC</u>: Carbon Copy: /CC to send a copy of your email to others

    <u>Subject</u>: In this field type in the subject of your email message

    <u>Body</u>: Below the subject; type in your message

    <u>Format</u>: Select your text in the 'Body' & Format>Font & select your style, /ok

    <u>Spell Check</u>: / (b) or press F7 <u>Attachment</u>: – To send email with attachments i.e. Pictures, Word, Excel etc.: >*toolbar* /Paperclip (b) & browse to find your file & once found //it to add it to your email.

        <u>Save Attachment</u>: If someone sends you an email with an attached message: //the attached *icon* to open it and select Save & browse to a folder you'd like it saved in & /Save

    <u>Send</u>: To send email on *toolbar* /Send (b)

    <u>Sent</u>: To view all emails you've ever sent: View>Folder List & /Sent Items *folder.*

Signatures – To send every email with a signature in its body i.e. Your Name with your title *President*: *menu* Tools>Options, /Mail Format *tab*, /Signature (b), /New & type in your name, /Next & use Font & create your own… /Finish, /ok, /ok. (/Inbox, /New message & notice your new signature in body of message)

    <u>Block</u>: To keep block signature from being added to every new message: Back in Outlook *menu*: >Tools>Options, /Mail Format *tab*, /Signatures for new message *arrow*>None, /ok

    <u>Change</u>: If you created more than one signature you can change your sig in an email: In your new email Right / your signature>your new sig.

Votes – To send out emails with voting buttons i.e. Accept or Reject to have pizza at your party instead of doughnuts: In Mail message on *toolbar* /Options (b) & /Use Voting Button's *arrow*>your choice (Note: once choice is added you can edit the default text and add more choices by using semi-colon i.e. Doughnuts; Salad; New Car! And doing this will create buttons in upper left corner of email that can be seen only by recipient). Address & send email & wait, & then you must open up each response or the following won't work…

    <u>Tally Votes</u>: View>Folder List, /Sent Items & open email that was sent to all voters & / its Tracking *tab* & info bar displays the tally of your votes.

Reply – To reply to an email: Open the email & /Reply (b) & type your message then /Send (Note: if someone sent you an email with an attachment, the attachment won't be sent back in your reply, but you must use Forward and re-enter their email address in To *field*).

Reply To All – When someone sends you an email and has Carbon Copied it to others you can reply to ALL of them including those CC'd: Open the email & /Reply To All (b) & type your message & /Send.

Forward – To pass an email with its original attachments on to someone else: Open email *toolbar* /Forward (b) & in To *field* enter email addresses to forward the message to.

Inbox – Contains all mail received & to receive email on *toolbar* /Send Receive (b)
    <u>Sort</u>: To sort emails: By Date /Received *bar* & /it again to sort again either ascending or descending.
    <u>Mark Unread</u>: To mark email as unread after you opened it and want a reminder to read it again: Right / email>"mark as unread."
    <u>Deleted Mail</u>: Goes to Recycle Bin folder and sits there until deleted a 2nd time. To delete mail permanently or 2nd time: On Outlook bar /Recycle *folder* & delete the email.

Recall – To recall a sent message, and the message can only be recalled if:
    1. It has not been opened.
    2. It has not been moved out of Inbox.
    3. The receiver is running Outlook and is logged in.
View>Folder List, /Sent Items *folder* & in that *folder* //a sent message & in its *menu* select Actions>Recall This Message… /ok. /Inbox & //the email reply for answer.

Appointment – To create a calendar appointment & using a pop-up reminder (only works if Outlook is running): Outlook *bar* /Calendar, *toolbar* /New & fill in the info.
    <u>Recurring</u>: An appointment that occurs every Monday for example. Open your appointment, & on *toolbar* /Recurrence (b) & select appropriate times.
    <u>Active</u>: To see Active Appointments: View>Current View>Active Appointments
    <u>Categorize</u>: To assign Appointment a Category: Open your appointment & in lower-right /Categories (b) & check one. To view your appointments by Categories: Close out of your appointment & *menu* View>Current View>By Category (Note: to back to original view: View>Current View>Day/Week/Month).

Event – Is nothing more than an Appointment scheduled all day: Open your appointment & check Event *box* (Note: you can schedule other appointments or more events that same day!)

Meeting – Is an appointment that others are invited to Accept or reject, hence a meeting: Open an appointment & on its *toolbar* /Invite Attendees (b), /To (b) to select email or type them in using a semi-colon as a separator & /Send when done
    <u>Replying</u>: 1st open your 'Invite' *mail* & on its *toolbar* /Calendar (b) to immediately view the requested date (The requested date will show what sender has blocked out and if you have something scheduled at the same time then it will show both, crunched in same time. Outlook will allow in this case over scheduling) and see if you have it available, then /Accept or Reject (Note: If you /Accept or Tentative the meeting will automatically be entered into your calendar after you hit Send (b))

Mouse click /	To >	Task Pane <u>TP</u>:
Button (b)	Enter or Return®	

Track Replies: To tally the yeahs & the nays: Remember, to track you MUST first open all emails. Go to Calendar & //on meeting you invited others to attend & /Tracking *tab*

Propose Time: Instead of accepting a meeting request; reply with a suggested time: Open Meeting request email & /Propose New Time & select a time & /Send (Now your new proposed meeting is scheduled on your calendar)

Update Meeting: If you have to make a change to the times of your meeting you can send an email update: In Calendar //a Meeting that has already been scheduled & /Scheduling *tab* and find its original time and drag its start & finish lines to different times...& on *toolbar* /Send Update

Cancel Meeting: – 1st delete a meeting from your Calendar & select 'Send Cancellation & Delete Meeting,' /ok & if you'd like type a message & /Send (Note: when recipient opens your message, a prompt to delete meeting from his calendar only if previously accepted)

Contacts – Your very own personal address book: In Outlook *bar* /Contacts & /New (b).

Category: To put your contacts into viewable categories: Open a contact & in lower-right *corner* /Categories (b) & select one OR /Master Category & type in your new category & /Save & Close

Phone List View: A nice simple way of viewing your categorized contacts: View>Current View> Phone List

Find: On *toolbar* /in Find Contact *box* & type in a name & hit®.

Tasks – To create a to-do *list*: In Outlook *bar* /Task (b) & in upper-right *corner* /New (b) (Note: REMINDERS will only pop-up as long as you have Outlook opened)

Check off Task: You can check the check box of the task when it's completed and it crosses a line through it to show completed, but you'll have to delete it yourself if you want it removed from your Task list.

Mini TaskPad: Note: in Calendar View on *toolbar* you can only select Day or Work Week to view the mini TaskPad in lower-right hand *corner.* To view only Tasks due today Right /blank area of TaskPad>TaskPad View>Today's Task.

Folder List – To view, organize, create and sort all of your folders at once: View>Folder List Right / Outlook Today (personal folder)>New & type in a name & hit®, /No. (Note: You can / & drag email from right window to a folder in your Folder *list*).

Find Messages – To find emails with keywords in one or more folders: Be sure you're in the Inbox & on *toolbar* /Find (b) & in Look for *field* type your *keyword*, /Search in *arrow*>folders you'd like searched, /Find Now (b)... When finished /Clear.

Notes – Act like real sticky notes only they're virtual and you can put them almost anywhere on your computer as little reminders: In Outlook *bar* /Notes, /New (b) & type in your note & /it's X to close it & it will show in Notes *folder*...(Note: You can also / & drag Note to your Desktop to store it there too).

Detail View: A different way to view your notes: View>Current View>Notes List

Assign to a Contact: In Notes, Open a Note & /its upper-left *corner* icon>Contacts & assign a contact & close Note. (Open that contact & /Activities *tab* to see Note has been assigned)

Assign to a Category: Open a Note & /its upper-left corne>Categories. To view Notes by their categories: On *menu* View>Current View>By Category.

Stationary – Applying background colors or themes to your emails: /Inbox, Tools>Options, /Mail Format *tab*, /Compose in this message format *arrow*>HTML, /Stationary Picker (b) & select any stationary & /ok. /New Email (b) (New Mail Message) to see your new background.

Customize: Tools>Options, /Mail Format *tab*, /Stationary Picker (b), /New (b) & type in a **name**, /Next & select Color & /its *arrow*>any color to be applied as the background to your emails, /ok, /ok, /Fonts (b) & /Choose Font (b) for "When composing a new message," & choose some styles for your text that will stand out from the background color you chose…, /ok & select "Always use my fonts," /ok, /ok. /New Email (b) to see your customized stationary. Change text back to black: Tools> Options, /Mail Format *tab*, /Use this stationary by default *arrow*>None, /Fonts (b), to Regular, size>10, color>Auto, /ok, /ok, /ok.

Message Formats – format in which an email is composed and or read in i.e. plain text, rich text or HTML:
1. Plain Text – messages won't look distorted to those who receive your email in their program that can't handle formatted messages
2. Rich Text – messages allow you to format text of your email to look nicer
3. HTML – or Hypertext Markup Language messages allow you to send emails with pictures as a background that can contain hyperlinks to
Tools>Options, /Mail Format *tab*, /Compose in this message format *arrow*>a message format, /ok.

Message Settings – used only to tell receiver how to treat the message that will be displayed when received and nothing more i.e.
1. Importance – Low, Normal & High
2. Sensitivity – Normal, Personal, Private & Confidential and is displayed after user opens up email, up top, on the InfoBar.
/New Email (b), /Options (b) & select your 'Importance' and/or 'Sensitivity' options.

Email Delivery Options – how, when and where to send your email: /New Email (b), on *toolbar* /Options (b) and…
1. Forward Replies – Type in an email address in the 'Have replies sent to:' *field* (if you want to forward to more than one address, use semicolons ; between each address).

Mouse click / Button (b)	To > Enter or Return®	Task Pane TP:

2. <u>Save Copies of Messages</u> – /Browse (b) for 'Save sent message to:' and select a folder you would rather save copies of your sent messages to rather than the default 'Sent Items' *folder*.

3. <u>Delay Mail Delivery</u> – check 'Do not deliver before:' & /corresponding *arrows*>date and time. The mail sits in your 'Outbox' until your specified time, and from that point forward you can /Send & Receive (b) and the mail will be sent then, or if your email sends out automatically, again, it will only be sent at the time you specified (Note: to send and receive automatically: Tools>Options, /Mail Setup *tab*, /Send Receive (b) & check 'Schedule an automatic Send Receive every…' & select your minutes. Be sure to leave computer on & outlook open to send automatically).

4. <u>Expire Your Messages</u> – messages that aren't read within a certain time will have a line through them to show it has expired. Great for emails you send with special, limited time offers. Check 'Expires after:' & /corresponding *arrows*>date and time.

Forwarding Contacts or Notes – to other people to have a copy of in their Outlook: In your Contacts *folder* Right/ a contact>Forward & enter an email address & /Send.

<u>Receiver</u>: The receiver of your contact opens their email, //the Contact & /Save & Close to save it to their Contacts (Note: if you receiver already has a copy of a contact, then a message will pop warning them of a duplicate).

<u>Notes</u>: Forwarding note follow same steps (concept) above…

<u>Export Contacts To Excel</u>: File>Import & Export, select 'Export to a file' /Next, select 'Microsoft Excel' /Next, /Contacts *folder*, /Next & /Browse (b) & type a **File name** of the Excel workbook you're exporting to select a place to save it, /Next, /Map Custom Fields (b), /Clear Map (b), from the left *pane* drag any *titles* you want exported over to the right *pane*…, /ok, /Finish.

<u>Link Contacts To Files</u> – in Outlook *folders* (like Inbox for emails) or other files outside of Outlook (Note: linking to outside files like Word documents, a shortcut of that file will be placed of your doc in Journal entry), so when you open up your contact you can /Activities *tab* and have one-stop- shopping access to those linked items: Open up a contact & from *menu* Action>Link> Items (or Files) and select a *folder*, then select the item(s) in that folder (use Ctrl *key* to select multiple items), /ok. <u>Use Links</u>: In your contact /Activities *tab*…

Distribution List – is a group of emails you entered under one word or phrase, and that one word (or phrase) acts as one email address to those groups of emails. Type that word in To: *field*, and it will send your email to everyone you entered into that particular Distribution List: On *toolbar* /Address (b), /New Entry (b) & select 'New Distribution List,' /ok, & type a **name** for this group of emails, /Select Members (b), & using Ctrl *key* select a few contacts & /Members (b) /ok (Note: you won't have a list if you haven't created any contacts with email addresses, but you don't have to create a new contact to arbitrarily add an email, simple /Add New (b)), /Save & Close & Close book. /New email (b), /To (b), select & select the **name** of your Distribution list, /To (b), /ok & type your email & /Send when done.

Assign Task – When you assign a task to someone it gets deleted from your Task *list* & is assigned – if accepted, but if not then upon return of declined task you have the option to put it back into your Task *list*. After you /Assign (b), notice: half way down you can check a couple of boxes... One to check is that anytime the person updates task or completes it, you'll get an email that will update the task in your task pad or mark it complete, but only after you open their email): /New Task (b), /Assign Task (b), in To: field type **email** & fill out task with due date..., /Send.

Receiver: In Inbox, open the "Task Request:" & on *toolbar* /Accept (b) (or Decline (b)) & select 'Edit the Response before sending, /ok & type in a **message**, /Send

Sending Task Updates: to send the person who assigned you the task update(s).

Receiver: In your Tasks, //The task you were accepted, /Status *arrow*>a message of your choice & for %Complete type a #, in message box above original message type in a desired **message**, /Detail *tab* & note the "Update List" has the email address of the person this will update to once you save this task with the changes.

Sender: When you open a "Declined Task" there's a new (b), "Return Task to List," that you can /on to put the declined task back into your Tasks.

Assigning Defaults: programming all your assigned tasks to have the same default settings: Tools> Options, /Task Options (b) & check or uncheck desired options, /ok

Sharing: to share a task without assigning it to anyone: Right / a task>Forward, enter **email**, >/Send.

Receiver: Opens the task sent via email & File>Copy to folder, /Tasks, /ok.

Multiple Message Sorting – Sorting messages with multiple criteria: Underneath any emails you have in your message window, in a blank area, Right / a white, blank area>Sort, /Sort items by *arrow*>any choice, /Then by *arrow*>a second choice, /ok.

Categorizing Messages – Assigning messages into categories can make the search through hundreds of emails easier if the user could narrow down the search to related groups: Right / an email>Categories, & check a category, /ok...

Group By Categories: Right / a blank area in your message window>Group By, / Group items by *arrow*>Categories.

Ungroup: Right / a blank area>Group By, /Clear All (b).

Filtering Messages – hiding unwanted messages for a time in a specific folder: Right / a blank area of message *window*>Filter & in From *field* type in an **email** address, /ok (Note: below Toolbar, the status reads (Filter Applied).

Remove Filter: Right / a blank area of message *window*>Filter, /Clear All (b), /ok.

Coloring Messages – applying specific colors to emails sent from certain people: On *toolbar* / Organize (b), & select 'Using Colors,' /Apply Color & the color red (default color) will be applied to the email (and ALL emails from same sender, and ONLY in your Inbox) you have selected below in the Inbox's message window.

Mouse click /	To >	Task Pane <u>TP</u>:
Button (b)	Enter or Return®	

Advance Find – Tools>Advance Find & type a **name** to find, to find those with attachments, from here, you can /In *arrow*>subject field and message body, More choices *tab* & check 'Only items with' one or more attachments & /Find Now (b) (note: as long as your Advance Find is open and you receive another message with same criteria it will show up in your Advance Find as well as your Inbox).

Rules Wizard – Creating criteria for incoming messages, like email with a specific word(s) to move to a designated folder: Tools>Rules Wizard (have to be in Inbox to find Rules Wiz in Tools *menu*), /New (b), to use a template, /Next, select 'with specific words in the subject' & uncheck 'from people or distribution list' & in bottom *pane* /Specified Words *link* & type a **word**, /Add (to add more words, type them & keep /Add), /ok, /Next & in bottom *pane* /Specified Folder *link* & select 'Personal Folders' /New (b) (to create a new folder to move the emails with certain words to) & type a **name** for your folder, /ok, /no, /ok (Note: selecting 'Personal Folders' puts your new folder at the same level of all the subfolders, like the Inbox when you view your new folder from View>Folder List) /Next, & select 'except if the subject or body contains specific...' & in *pane* below /Specific Words *link* & type a **word** that is an exception and will keep the email from being move to your new folder, /Add, /Next & check 'Run this rule now...' /Finish, /ok. To see your new folder: View>Folder List.

Calendar Conditional Formatting – To apply specific colors to appointments that contain specific word(s): In the Calendar, on *toolbar*, /Calendar Coloring (b)>Automatic Formatting, /Add & type a **name** for this condition, /Label *arrow*>a color (or category), /Condition (b), & type **word** or phrase, /ok, /ok.

Customize Toolbar – Adding or removing (b)s from your toolbars: /Inbox & at very end of *toolbar*, /Toolbar Options *arrow*>Add or Remove Buttons>Standard & check to add a (b) or uncheck to remove one, or Right /Toolbar>Customize, /Commands *tab* & under 'Categories' select 'Edit' & in 'Commands' you can / & drag 'Cut' or 'Copy' next to the rest of the (b)s on the *toolbar* above (Note: you can't drag new (b)s anywhere they have to be next to the current (b)s on your toolbar). Remove: Right /Toolbar>Customize & with the Customize box open you can / & drag (b) off the toolbar into a blank area and the (b) will disappear.

> New Toolbar – Right /*toolbar*>Customize, /Toolbars *tab*, /New & type a **name**, /ok, >/Commands *tab* & add any (b)s from under the 'Commands' *pane* by / & drag to your new toolbar.

> > Delete Toolbar: Right /*Toolbar*>Customize, /Toolbar *tab*, select your toolbar & /Delete (b).

> New Menu – Right /*toolbar*>Customize, /Commands *tab*, under 'Categories' select 'New Menu,' & under 'Commands' / & drag 'New Menu' above *toolbar* & next to the Help *menu* & Right /it & in Name type a **name**. Then drag a few commands like Copy or Paste to it... To delete the menu, / & drag it off to a blank area (Note: the Customize box has to be open to remove menus or make any changes to the toolbars).

Customize Inbox View – add or remove column headers in message view of Inbox: /Inbox, Right /inside a blank part of the message *window*>Show Fields, //a desired field from left *pane*, /ok (Note: the field is added, but you can / & drag the column field bar to move it ahead or behind other column headers). To remove *header*, / & drag the *header* off into a blank area & let go of mouse.

2 Time Zones – adding an additional time zone to compare & contrast with current one. Only displays in calendars view's 1 day or 5 Work Week: Go to Calendar, on *toolbar* /1 Day (b) & Right / the gray bar that displays the times by hours>Change Time Zone & check 'Show additional time zone' & choose your zone & enter a **label** for it, /ok. To remove right / & uncheck it.

Archiving – moves what you choose from: Outlook's messages, tasks, contacts and notes (to move items that you still need out of your hair, but don't want to delete and possibly import back in later if needed), into another location that you can import later is so desired: File>Archive & select Inbox, /Archive items older than *arrow*>a date that you would like to start archiving when reached, /ok.

To Import Back: File>Import and Export & select 'Import from another program or file,' >/Next & select 'Personal Folder File (.pst)' /Next, /Browse (b) & // Archive.pst, /Next & select Inbox (if you want to restore inbox items only), /Finish (Note: import will copy from the archive folder your data, and your originals will still be in the archive file).

Automatic Archiving: Right /Inbox>Properties, /AutoArchive *tab*

Mouse click /	To >	Task Pane TP:
Button (b)	Enter or Return®	

PowerPoint

Templates – Default slides that are preformatted you can use: Open PowerPoint & in *Task Pane*: /General Templates, /Design Templates *tab* & select a template, /ok.

Add Slides – To add more slides to your presentation: *Task Pane*: find a desired slide & with your mouse hover over it & then /its *arrow*>Insert New Slide.

Change Slides – To changes a slide's layout: 1ˢᵗ make sure your slide is selected and in *Task Pane*: hover over a desired slide with your mouse & /its *arrow*>Apply to Selected Slides (Note: if your *Task Pane* is not visible then in *menu* View>Task Pane, OR Right/ your slide>Slide Layout).

Edit Slide View – By default when you open PowePoint the left side of your screen has an Edit *bar* with 2 *tabs* (One Slide & other Outline), and Slide is selected. (Note: you can close this bar, but if you want it back in *menu* View>Normal Restore Panes).

Edit Outline View – This view is to edit the text of your slides in an easier to manage "Outline View": On left side of screen /Outline *tab*.

Clip Art – To insert Clip Art on a slide 1 of 2 ways:
1. From the *Task Pane* find a Slide that has 'Content' & insert it. On that slide /Insert Clip Art (b) & select your art & /ok
2. Or insert a blank slide & Insert>Picture>Clipart & in *Task Pane* /Clip Organizer *link* & in Collection *list* expand 'Office Collections' then find your *folder* and in it Right/ your art>Copy, minimize the 'Organizer' & Right /on your slide>Paste.

Bullets – To add bullets or a numbered list slide to your presentation add a slide with Text: On *Task Pane* find a Text slide (has bullets in it, just add text) & insert it
Moving: To rearrange the order of your bullets or slides for that matter:
1. Go to 'Outline' *view* (Left *pane* /Outline *tab*), then simply /on the bullet (or slide), hold and drag it to a new desired location.
2. Or select a bullet & on the Outline *toolbar* /Move up or down *arrows*
3. Another way to add bullets to a blank slide is by adding a Text Box: On Drawing *toolbar* /Text Box (b) & / & drag to draw a Text Box on you slide, then up on *toolbar* /Bullets (b) & you've added bullets to your floating Text Box.
Promote/Demote: You can turn demote slides into bullets or promote bullets into slides to help manage your outline: Bring up Outlining *toolbar* (View>Toolbars>Outlining), select a bullet & on *toolbar* /Promote (b) (it's a green arrow pointing left) to promote a bullet to a slide. To demote it /Demote (b).

AutoShapes – Adding shapes to your presentation with text in them: i.e. On Drawing *toolbar* /AutoShapes (b)>Stars & Banners & select a Star & use shift key to / & drag a proportional star on your slide & type in some text.

Word Wrap: To wrap your text in the shape: Right/ your shape>Format AutoShape, /Text Box *tab* & check Word Wrap, /ok

Change AutoShapes: To change your shape: Select the shape & on Drawing *toolbar* /Draw> Change AutoShape>Whatever you want

Color: To add color to a shape or object: Select shape & on Drawing *toolbar* /Fill Color (or Paint Bucket) *arrow*>a color or>'Fill Effects' & choose between the 4 *tabs*: Gradient (2 colors or more), Texture (Marble, oak, cork…), Pattern (dots & lines) and Picture (You can insert a picture into your shape by /Select Picture (b) & browsing & //on a picture you want to insert.

WordArt – Basically it's text with attitude or flair that you can add 3-D effects to: Select your text you want converted to WordArt & on Drawing *toolbar* /WordArt (b), //any of your choice, /ok. To resize your WordArt simply / & drag one of its white handles around the WordArt out or in. To add 3-D effects /the last button on Drawing *toolbar*>3-D Settings & use that toolbar to add effects like lighting, depth & rotations.

Edit: To edit WordArt, select your Art & on WordArt *toolbar*, /Edit Text (b). To change your Art, on WordArt *toolbar*, /WordArt Shape (b) (abc) and select any shape, /ok

Drawing toolbar – Used to make shapes: To make a circle /on oval & hold Shift key & / & drag to create a perfect circle. For square /Rectangle & use Shift key too.

Org Chart – To insert an Organization Chart in your slide: In *Task Pane* scroll to the bottom & look for a slide that looks like a layout of an Org chart, then hover over the slide & /its *arrow*> Insert New Slide. //on center of slide & /ok. /on the boxes & type in the names of your of your coworkers and their titles.

Add Boxes: Select an employee box you'd like to add a subordinate, coworker or assistant to & on Organization Chart *toolbar* /Insert Shape *arrow*>any of the 3.

Delete Boxes: /on a box & press delete

Move Boxes: Hover your pointer over border of a box you want moved till you see 4-way arrow and / & drag it over the box of another employee to change the structure.

Style: To change the style of your Org Chart: On Chart *toolbar* /AutoFormat (b), select one & /Apply

Table – Creates many cells to place and organize text in: *Menu* Insert>Table & type in your rows & columns, /ok. Basically, all you need to add or remove rows, columns and cells

Mouse click /	To >	Task Pane TP:
Button (b)	Enter or Return®	

45

is on the Tables & Borders *toolbar* using its Table (b) *arrow* or Right/a cell in your table>delete or insert.

Borders & Fill: To add color to your table's borders or shading to cells: 1ˢᵗ select your table or a cell you want to apply color & Right/the cell or *fuzzy* border of your table to have color or borders applied to the whole table>Borders & Fill… (Note: if you are trying to apply a border style or color to a border around your table or cell be sure that after you select your style you don't close your Borders & Fill *window* until you apply that Border style by clicking in its *preview box area*).

Chart – Inserting a Bar Chart: On *Task Pane* scroll to its bottom & hover over the Chart *template* & /its *arrow*>Insert New Slide. On your new slide //where its says "Double click to add chart" (Brings up a default Chart template that you can delete and add your own data in its floating Datasheet)

Datasheet: When you're done entering your data in the Datasheet don't /the X or its Close *window* (b) (Because if you do, you can't //the chart later to easily open the Datasheet to make data changes to your chart). Instead /in a blank area off of your slide to close your Datasheet & your Chart's edit mode. If you did /the X, the only other way to bring up your Datasheet is a little *tricky* because you have to Right/ the white space between the chart & it's outer *fuzzy* border>Datasheet

Chart Type: To change your chart to a pie, column, doughnut chart, etc. Make sure you chart is in edit mode (Datasheet is showing. You can //on your chart to open Datasheet) & *menu* Chart> Chart Type & select any chart, /ok

Chart Options: To add a Title to your chart, Data Labels etc.: Chart>Chart Options.

Formatting Text – To change the formatting of your text: Select your text *menu* Format>Font

Line/Bullet Spacing: To create single or double spaced bullets or lines in your text: 1ˢᵗ select your text or bullets & Format>Line Spacing & increase spacing>your choice, /ok

Indents: To indent your text or bullets all at once: 1ˢᵗ make sure your horizontal ruler is showing by View>Ruler, then select your bullets or text. On H-ruler grab the 'Left Indent' *marker* (Looks like a little box under attached little triangle) & drag right an inch (Note: all sub-bullets or text are now indented).

Spell Check – To check your presentations spelling: Right /any misspelled word (underlined in red), or to check the whole presentation on *toolbar* /Spell Check (b).

Slide Sorter – A view to see ALL of your slides, arrange slide, add & see a preview of your slide transitions: View>Slide Sorter

Hide Slide: To hide a slide from displaying in your presentation without deleting it: /on a Slide & on *toolbar* /Hide Slide (b) (Note: your slide # has a gray box around it with a line through it to show it's hidden). To unhide select your slide & /the same (b) again to unhide.

Slide Order: If you'd like to rearrange your slides so let's say slide #3 becomes your first slide: /on Slide #3 & drag it before Slide 1

Transitions: In your presentation when you / from one slide to advance to the next; how that slide pops-up, >scrolls in etc. on the screen is called a transition: Right/ a slide>Slide

Transition & in *Task Pane* select a transition; you can also select your speed, a sound & whether to activate your slide on Mouse / or after so many seconds.

Custom Animation – To animate your bullets or text when they enter in your slide: Select your text (or a bullet), Slide Show>Custom Animation & in *Task Pane* /Add Effects (b)>Entrance>any effect.

Speaker Notes – Notes that can be added to each slide & printed out for the presenter to read during their presentation: View>Notes Page & under a slide type a few notes. To print note pages with accompanying slides: Cntrl+P, /Print What *arrow*>Notes Page.

Print handouts – To print slides for handouts in Black & White: View>Slide Sorter, on *toolbar* /Color Grayscale (b) (a red, green & blue (b))>Pure Black & White. File>Page Setup & under 'Notes, handouts & outline' select Landscape & you're ready to print.

Slide Show Shortcuts – Use following the when viewing your slides in:
★ Press 8® (to go to slide 8)
★ Press W (White out)
★ Press B (Black out), press B again & slide reappears
★ Right/slide>Pointer Options>Pen (press ESC to exit pointer)
★ Press E (erases)
★ Press N (next slide)
★ Right/>Pointer Options>Automatic & /slide
★ Press H (shows hidden slide)
★ Hold both mouse buttons down for about 2 seconds (computer beeps) and you go back to slide one.
★ Press F1 (view shortcuts).

Slide Master – If you want the same format or style on every slide all you need to do is change the Slide Master: *menu* View>Master>Slide Master, now make your changes & when finished /Close Master View (Note: every slide will have your changes made to your Master except any Title Slide, the Title Slide has its own Master…)

Title Slide Master – Generally, footers, slide #'s & logos don't appear on a Title slide except the Title of your Presentation: View>Master>Slide Master & on the Slide Master floating *toolbar* /Insert New Title Master (b), & make your changes i.e. select the text "Click to edit Master title style" & change its font to bold or red & it'll apply to all slides with a Title (But not the 'Title Slide') & when finished on *floating toolbar* /Close Master View (b)

Add Footer – To add a footer to all of your slides like dates, slide numbering & notes: View>Header and Footer (Note: you can apply changes to All Slides or just one you're currently working on).

Mouse click /	To >	Task Pane **TP:**
Button (b)	Enter or Return®	

Format Bullets – To change your bullets to a picture: Select your bullets on a slide & Right / them> Bullets & Numbering, /Picture (b) & /on a picture once, /ok.

Design Template – If you'd like to preserve your current presentation's style to base all of your future presentations off of, then save it as a Template: File>Save as, /File Type *arrow*>Design Template, & type a name for your Template, /Save.

New Presentation: Based on your saved template: 1ˢᵗ close out of your template you just saved, File>Close & then File>New, in *Task Pane*, /General Templates *link*, & //Your Presentation you saved as a template (Note: You now have a new presentation based off of your template you saved earlier, AND if you now save this new presentation is won't overwrite you template, but save it as another presentation).

Advanced Objects – Using Drawing toolbar for creating & manipulating object for a more visually festive presentation:

Group Objects: To group objects as one: /on your first object & then hold Shift key & /on the rest of your objects to select them. On *toolbar* /Draw>Group. To ungroup: /Draw> Ungroup.

Align Objects: To align objects perfectly within each other: Select all of your objects to align with in each other by using Shift key to help them, On *toolbar* /Draw>Align or Distribute>Align Center, then /Draw again>Align or Distribute> Align Middle.

Order: If some objects are hiding behind others and you want them in front: Right / the object you want in front>Order>Bring to Front.

Rotate: Most objects when selected have a green handle (circle) that when / & dragged will rotate your object freely.

Sound – You can attach a sound to an object on its entrance on any slide. (Note: this exercise will only work if you've already created an object & have assigned it an animated effect). Let's say I've created a circle to fly in & I want the sound of breaking glass upon its mouse clicked entrance: You DON'T need to select the circle, Insert>Movies & Sound>Sound From File & browse through your computer to find your sound, //it & /No (so it doesn't play automatically)...

Hide Sound: After your sound is inserted you'll see a little speaker representative of your sound. This speaker will also show in your presentation too, but if you only want to hear it & not see it: Drag sound icon off to side of the slide in the gray area.

Cue Sound: Select that speaker that you've dragged off to the slide's side & Slide Show>Custom Animation & in *Task Pane* (This next step is a bit tricky) your sound will have a Tan colored *title* & below it the actually sound itself. / & drag the 'actual sound itself' above its Tanned title (If done correctly that Tan *box* will disappear & that's important for almost ANY sound customization, and in this case to synchronize with other animated objects). *Task Pane* / on your sound to select it then / its *arrow*>Start with Previous (If your previous is the animated circle it will sound at the same time you / your mouse to execute that circle to fly in). Test it!

Animate Objects – You can have bullets or item list and any objects to come in on cue with a mouse / on any slide in your presentation: Right /on your object you want animated>Custom Animation, *Task Pane* /Add Effect>Entrance>any effect.

<u>Order of Effects</u>: If you have more than one object animated in a slide you can have them animate in a order you choose: Select your 1ˢᵗ object (or animated bullet) & *Task Pane* / & drag the object's animation *bar* in the order you'd like it to animate (Top of the animation list will be 1st)

Export to Word – You can export your presentation to Word for a printable version: File>Send>Word & select an option, /ok

    <u>Meeting Minder</u>: To take notes during a presentation which can are stored into the presentation/ slide itself: Start your Slide Show & Right / on any slide>Meeting Minder & type In a note or reminder &, /ok

    <u>Action Items</u>: If you have Microsoft's Outlook you could, during your presentation, schedule yourself a Task reminder: Start your Slide Show & Right / a slide>Meeting Minder & /Action Items *tab*. In Description, describe the nature of your task, enter the name of person assigned to it, /Add. / Schedule (b) & enter your reminder information OR /Export (b) & Export it to Word & schedule a Task at the same time.

Automatic Slideshow – To assign time to each slide that will automatically advance after its time expires: Go to Slide 1 & Slide Show>Rehearse Timings & watch the timer that pops up & / when you think enough time has been spent on that slide to stop the timer & advance to the next slide & the timer will start up again. Continue till you get to end of your show & /yes to save your timings for each slide.

    <u>Kiosk</u>: You could use your presentation in a kiosk & your Slide Show will go on a continuous loop until somebody hits ESC key: Slide Show>Set Up Show & /Browsed at a kiosk, /ok

    <u>Override Timings</u>: to have your presentation ignore your rehearsed timings without erasing them: Slide Show>Setup, under "Advance Slides" select 'manual', /ok.

Record Narration – To record your voice for each slide. Be sure you have a microphone hooked up to your computer: /Slide 1, Slide Show>Record Narration, /ok & speak into the mic what you want recorded & / to advance to the next slide do the same until the end of you presentation & you can then /yes to keep both time you spent on each slide & your narration.

Pack & Go Wiz – To save (includes your fonts), transport & use your presentation on a computer that doesn't have PowerPoint: File>Pack And Go, & follow steps in Wizard. When finished find the folder or disk you Packed your presentation too & note the files you'd use to install on the computer that doesn't have PowerPoint are the PNGsetup.exe & .PPZ files.

Presentation Wiz – Creating a presentation using AutoContent Wizard. To create a presentation that is non- linear (using links like a web page): File>New, in *Task Pane* /From AutoContent Wizard *link*, /Corporate & select 'Group Home Page,' /Next & select 'Web Presentation,' /Next & follow the rest of the Wizard. When finished Press F5 to start

Mouse click /	To >	Task Pane <u>TP</u>:
Button (b)	Enter or Return®	

Slide Show... (Note: On the left side of your presentation there are several titles that are actual links to other slides in your presentation. Also, note that you can change the design, formatting, text of this presentation to your design). To end show Press ESC *key.*

Create Links – To create your own links like you'd find on the internet, but to navigate your own slide show in a non-linear way: On slide type some text (or insert an object) that during your presentation when you / on it, you'd like to take you to another slide. Select that text (or your object) & Right / it>Hyperlink, /Place in this doc, & then select a slide you'd like this link to take you to when you / on it & /ok. Press F5 & test your *links!*

Outside Links: To link your slide to another web site: Select some text or an object on one of your slides & Right / it> Hyperlink & in Address *box* type www.disney.com, /ok

Action Settings: to hyperlink an object with specific actions: Select some picture on any slide: >Slide Show>Action Settings, /Hyperlink to, /Hyperlink *arrow*>Other File & browse and // another presentation, /ok.

Link Excel Chart – You can insert an Excel Chart you've created into a slide & have it linked so any changed you make in Excel will update that change in your slide: In a blank slide Insert>Object, /Create From File, /Browse>a place you have stored your Excel Chart & // it, check Link *box*, /ok.

Edit: If you //on your inserted Chart on your slide, this will open up you chart & data in Excel. (Note: if you have your Chart on 1 worksheet & the Chart's Data on a 2nd worksheet, if you make any changes to the data & save, you will no longer have a chart in your slide. This is because PowerPoint will link the last thing you did in Excel. So edit your Excel workbook, but as the last thing you do select your chart & /Save & then close Excel & since you touched your chart last then PowerPoint will display your Chart).

Publish to Web – To prepare your presentation for the Web: File>Save as Web Page, /Publish, /Change & type a name that you want others to see up in the blue Title bar when they visit your website, /ok, /Publish. (Note: Your web page will be stored in the same folder your Presentation was saved in).

Import Word – You can import a Word Outline into PowerPoint & it will convert it into slides & bullets accordingly: Insert>Slides From Outline, browse for your Word document & //it.

Word Outline: You must set up your Word document in outline format that PowerPoint can understand when you import it: 1st use for your headings styles, Heading 1, 2 or 3. In other words, Heading 1 style when inserted will each have their own slide & will be the title of those slides. Heading 2 will be inserted as bullets supportive to each Heading 1 title.

Review – You can have others people review you presentation & add their comments for you to later accept or reject them:

1. You: First, File>Save As, /File type *arrow*>Presentation for Review & rename your presentation & save it in a place where others can have access to it & /Save & File>X.

2. Reviewer: Second, the person(s) reviewing your presentation will open it up & / No (Not to merge changes with your original Presentation) and make any changes & when finished they will /Save & File>Close

3. <u>You</u>: When everyone has had a chance to review & make some changes, then you open it up & /Yes (to merge changes with your Original Presentation. When the changed presentation merges with the original the discrepancies will be displayed Slide-by-Slide for you to accept or reject). On the Reviewing *toolbar*, /Next Item & on each slide check to see if you like the changes & if so /All changes & to continue /Next Item, & accept or reject. When you're finished on Reviewing *toolbar* /End Review (b), /Yes & /Save. (Note: Your Original Presentation now has all changes made or none if you've rejected them, and now you can delete the copy of your presentation that you made available in step 2 for others to review).

Online Broadcast – To give a live presentation over the web with voice & visual (Also, can be recorded and viewed later on the web): Go to Slide 1, Slide Show>Online Broadcast>Schedule a Live Broadcast & type the name of your presentation, /Save, /Settings, /Browse to select a shared network directory *folder* & under "Video & Audio" select your choice, /ok. /Schedule (b) & enter the times & email address of all those to attend & when finished /Send (b) (Note: The recipient will receive your email & in it will be a hyperlink address that they can / on to open up the their internet & link right to the broadcast & all you have to do from your end is when it's time is: Open your Presentation & on *menu* Slide Show>Online Broadcast>Start Live Broadcast Now, /Yes & select your scheduled time & /Broadcast (b))

Shortcuts

<u>Ctrl+A</u> – Selects *all*.
<u>Ctrl+Z</u> – Undo any action
<u>Ctrl+Y</u> – Redo any undid action.
<u>Ctrl+S</u> – *Save*
<u>Ctrl+X</u> – *Cut*
<u>Ctrl+C</u> – *Copy*
<u>Ctrl+V</u> – *Paste*
<u>Ctrl+F</u> – *Find*
<u>Ctrl+P</u> – *Print*
<u>Ctrl+D</u> – Duplicate slide (Note: if an object is selected & you Ctrl+D it creates a duplicate of that object)
<u>F4</u> – Like Format Painter, *but* it only applies *most recent* action or formatting performed.
<u>F5</u> – View Slide Show
<u>F7</u> – Spell checks to entire presentation
<u>Cntrl+Home</u> – Puts cursor at the very beginning of the slides.
<u>Cntrl+End</u> – Puts cursor at the very end of the slides.
<u>Home</u> – Puts cursor at beginning of a line.
<u>End</u> – Puts cursor at end of line.
<u>Tab</u> – in Outline view will demote a slide to a bullet
<u>Shift+Tab</u> – in Outline view will promote a bullet to a slide
<u>Ctrl+/ & Drag</u> an object or some selected text to create a copy of it

Mouse click /	To >	Task Pane <u>TP:</u>
Button (b)	Enter or Return®	

Word

Word Wrap – When you type your document & get to the end of your first line you don't have to hit ® (unless you want a new paragraph), but keep typing & Word will wrap your text for you on the next line.

    <u>Hard Returns or Paragraphs</u>: When you hit Enter® on your keyboard Word creates a paragraph for each ®. This is fine if you need extra spacing, otherwise let Word Wrap your sentences for you, otherwise other more advanced techniques will become obsolete.

    <u>Soft Returns</u>: Shift+® (forces new line break without inserting a paragraph).

Show Hide Codes – To see the background codes of your document on *toolbar* /Show Hide (b) (and / this (b) again to turn off codes. Dots are spaces & arrows are tabs).

Save – When you Save your document for the 1st time Word performs a Save As! But after you Save it once all you have to do is on *toolbar* /Save (b) & it will save without any questions.

Save As – When a document is Saved for the 1st time Word will open a small window to ask you a couple of questions before it can save your document:

1. <u>Where</u> do you want to save your document: /Save in *arrow*>a place like a folder or your Desktop
2. <u>What</u> is the name of your document: In File Name *box* type in the name of your document & /Save

Also, if you have a document 5 pages long addressed to Sam & you want to create a copy for Cathy without having to retype it: File>Save As & rename the File Cathy. Now you'll have 2 documents: 1 for Sam & 1 for Cathy.

Smart Tags – Little boxes that pop up in certain places for particular reasons: For example if you type in a person's name & address in Word & hover over it with your pointer a *box* pops-up for you to / on to choose to enter this address as a Contact in Outlook XP Or to look up a map for this address on the internet! (Note, you have to have Microsoft's Outlook program installed on your computer to add it as a Contact)

    <u>Turn Off</u>: For some people all those little boxes (Smart Tags), can be annoying & turn them off: Tools> AutoCorrect Options, /Smart Tags *tab* & uncheck first box 'Label Text with Smart Tags,' /ok. Also, Tools>Options, /View *tab* & uncheck Smart tags.

AutoText – You can program Word to insert a phrase after you type its first 4 letters & hit ®: *Menu* Insert> AutoText>AutoText & type in your phrase, /Add, /ok. Now simply type in the first 4 letters of your phrase & watch for a yellow *tag* to appear, when it does hit ® to accept it.

    <u>Delete</u>: To delete your AutoText entry: Insert>AutoText>AutoText & type in the first letter of your phrase & look in the window below for your complete phase & select it & /Delete (b).

Help – If you have any questions you can type your question up on toolbar in the "Type a question for help" box & hit ®.

Selecting Text – There are several ways you can select a word or a sentence in your document for any editing purposes:

<u>Word by word</u>: Cntrl+Shift+left or right arrows on *keyboard*

<u>Line</u>: Cntrl+/Line Or Move pointer into your left margin of your document until it point's right then /

<u>Word</u>: // a word

<u>Paragraph</u>: ///

<u>All</u>: Cntrl+A

<u>Multiple Selections</u>: Use Ctrl *key* to select multiple text in scattered through your document.

<u>Point>Home or End</u>: You can select everything from wherever your cursors flashing> Home or End of your Document when you Cntrl+Shift+End or Cntrl+Shift+Home.

Undo Redo – If you made a recent mistake in your document you can Undo it: On *toolbar* /Undo (b) (a blue *arrow*) AND if you Undo something & decide you want it back /Redo (b) on *toolbar*.

Move Text – You can move your text around 3 ways:

1. <u>Drag</u>: Select your text & with your mouse / & drag it to wherever you want on your document.
2. <u>Cut</u>: Select your text & on *toolbar* /Cut (b) (your text will be cut to a clipboard that isn't visible to you unless you make it so), & / in a place with your mouse on your document & /Paste (b).

Copy Text – Select your text & on *toolbar* /Copy (b) & put your cursor in another part of your document & /Paste (b) OR you can select your text & hold Ctrl *key* & drag with your mouse to another part of your document.

Clipboard – When you Cut or Copy your text it goes to the clipboard & stays there until you paste it on your document. In fact, you can have cut or copied up to 24 items on your clipboard to later paste individually, in the order you'd like, in another part of your document OR another document all together: Edit>Office Clipboard. You'll see the *Task Pane* is your clipboard & anything you cut or copy will display in it up to 24 items. After which you can then Right / any item in your clip board> Paste.

Format Text – To change the format of your text to bold, italic & color: 1st Select your text & *menu* Format>Font & choose your formats here & /ok when you're finished.

Mouse click / Button (b)	To > Enter or Return®	Task Pane <u>TP</u>:

<u>Quick Format</u>: To apply the most recent 'single' format style to other text: For example, select some text & on *toolbar* /Bold (b) & now select some different text & press F4. (Again, this quick format only applies a single, most recently applied format. For copying multiple formats to apply to other text see Format Painter…)

<u>Format Painter</u>: Used to copy multiple format styles from on text to apply to other text: Find some text that has some formatting (like bold, red, size 24 etc.) you'd like to copy & apply it to some plain text & / in it so your cursor is flashing in the middle of it & on *toolbar* /Format Painter (b) & drag your brush over the plain text to paint or apply your format to it. (Note: Your Format Painter will only paint once & disappear, unless you //the brush, and it won't stop painting until you / once the brush again or hit ESC *key*).

Highlighter – Word has a highlighter you can use to highlight your text: On Formatting *toolbar* /the tiny down *arrow* to the right of the Highlighter (b)>select a highlighter color. Then / & drag over text you'd like highlighted (To stop highlighting / Highlighter (b) or press ESC *key*).

Paragraph Alignment – You can have your paragraphs aligned on your document with the Alignment (b)s on Formatting *toolbar*: / in a paragraph you want aligned & on *toolbar* select your alignment (b) & /it i.e. /Left aligned (b).

Line Spacing – For single, double or customized line spacing on your document: Select the text you want to change its spacing & on Formatting *toolbar* /the *arrow* to right of Line Spacing (b)>your choice spacing OR use shortcut keys Ctrl+1 is single; Ctrl+2 is double; Ctrl+5 is 1 ½ spacing.

Borders & Shading – You can put colored borders around your text & fill-in that bordered text with a shade: Select some text you want a border around & Format>Borders & Shading, /Borders *tab* & select a Style, Color & Width & /Apply to *arrow*>your paragraph or text; Now for shade /Shading *tab* & select a color & /its Apply to *arrow*>text or paragraph, /ok.

Tab Stops – When you hit your Tab key Word tabs 5-spaces (default), but you can set your Tab to stop more or less with your text aligned to the left, right or center of those stop: For example, let's say we want to create 3 columns of information like so – **Books   Author   Cost** On a blank page type *Books*. / on your Vertical Ruler to deselect your page (if you don't have a Vertical Ruler then View>Print Layout). Then just above your Vertical Ruler you'll see a little box with an L shape in it, hover over this with your mouse & it ought to say "Left Tab." If you / on it, it will toggle to other Tabs. Toggle to the 'Center Tab.' Then with your mouse /on the Horizontal Ruler's 3-inch mark (the Center Stop ought to be inserted, if not try again). Hit the Tab *key* & it will tab to 3-inch Tab stop & then type **Author** (Note: Author is centering off of 3-inch Center Tab stop). Toggle through the other Tabs to find Right tab & / on Horizontal Ruler's 5 ½ inch mark. Hit Tab *key* & type **Cost** (Cost to the left of your Right Align Stop). (Note: when you hit ® and you go to a new blank line the tab stops you set will still be there on your horizontal Ruler; so all you have to do is hit your Tab *key*)

<u>Clear Tabs</u>: To remove Tab stops from your page you: If you wanted to remove one stop like the Center stop; simple / & drag that stop off of your Horizontal Ruler. Another quick way is Format>Tabs, /Clear All (b), /ok

<u>Tab Leaders</u>: are dots, dashes or lines that track the reader's eyes from one Tab stop to the next i.e.

>**Books   Author   Cost**

1st select ALL your lines that you want Tab Leaders in (Note: Tab Leaders are based on Tab stops, so be sure to have your Tab stops in place before inserting your Tab Leaders!), then Format>Tabs & in the big box you ought to see a list of your 2 Tab stops. Select the 1st 3-inch stop & under 'Leader' select #2, /Set (b). Now select 2nd 5 ½ inch stop & under 'Leader' select #2, /Set (b), /ok

<u>Clear Leaders</u>: To clear your leaders: Select all your lines of text that have Leaders & Format>Tabs & select your 1st Tab stop 3-inch & under 'Leader' select #1, /Set (b), /ok.

Indent Markers – Up in the Horizontal Ruler on the left side are 3 small Indent Markers that can help adjust your paragraph's indentations:

<u>First Line Indent</u>: top triangle; when you / & drag, right indents 1st line of paragraph.

<u>Hanging Indent</u>: middle triangle; when you / & drag it right indents all but 1st line of your paragraph.

<u>Left Indent</u>: bottom rectangle; when you / & drag it right indents the whole paragraph.

Bullets – Used in numbering paragraphs or to create lists: Format>Bullets and Numbering

<u>Spacing</u>: To create spacing between your List Items or Bullets: Select your bullets & Format> Paragraph, /Indents & Spacing *tab*, under 'Spacing' change 'After' from 0 pt>6pt, /ok.

Table – Creates many cells to place and organize text in: *Menu* Table>Insert>Table & type in your rows & columns, /ok. Basically, all you need to add or remove rows, columns and cells is in the Table *Menu* or Right / a cell or column in your table>delete or insert (Note: if you're in the last row of your table & you need another row, simply tab till you see your new row).

<u>Tab</u>: To advance from one cell to the next in your table

<u>Shift+Tab</u>: To tab backwards in your table

<u>Alt+Tab</u>: To actually tab in the cell without advancing to another cell

<u>Borders & Shading</u>: To add color to your table's borders or shading to cells: 1st select your table or cell you want to apply color & Right / it>Borders & Shading... Make sure that the Apply to *arrow* on Border or Shading *tab* displays what you want to apply you formatting to i.e. cell or table.

<u>Table AutoFormat</u> – To use Word's preformatted tables on your table: 1st /inside your table & *Menu* Table>Table AutoFormat & select, /ok.

Mouse click /	To >	Task Pane <u>TP</u>:
Button (b)	Enter or Return®	

Template – An original document you don't want changed, but the original can be duplicated whereas any modifications are made upon the duplicates without affecting the original (Template): Create your document, memo, letter or whatever you want to save as a template & *Menu* File>Save As, /Save as type *arrow*>Document Template.dot, & type name of your template, /Save & File>Close. To extract a copy of your template: *Menu* File>New, in *Task Pane*, /General Templates *link*, /General *tab*, //on your Template. Delete: *Menu* File>New, in *Task Pane*, /General Templates *link*, /General *tab*, Right / your Template>Delete

> Fax: To create a fax document using the Fax Wizard Template: File>New, in *Task Pane* /General Templates *link*, /Letters & Faxes *tab* & //Fax Wizard & follow the prompts.

Thesaurus – Right / any word>Synonyms.

Find & Replace – To find text or find & replace text: Edit>Find & type your text in Find *box* & check 'Highlight All Items Found' & (Note: /More (b) to find case sensitive words etc.) /Find (b). To replace /Replace *tab* & in Replace *box* type what you'd like replaced & /Replace All (b).

Spell Check – Right/ on any misspelled word or on *toolbar* /Spell Check (b). If you find while Spell Checking & find you need to make changes you can do it right in the Spell Check i.e. delete words or add some, but be sure to /Change (b) to accept your changes & for Spell Check to move to next error.

Header & Footer – 1st be sure you're in Print Layout *view* (View>Print Layout. You have to be in this view to see, add or edit your Headers & Footers) View>Header and Footer, scroll to the top of any page & in the Header *box* or Footer type in some text or insert a logo you'd like duplicated in that same spot on every page. When finished on the Header & Footer *toolbar* /Close (b).

Page Numbering – Insert>Page Numbers, select a format, /ok.

Page Breaks – To insert a new page you can either:
  1. Ctrl+® or
  2. Insert>Break, select 'Next Page' /ok (Note: This will not only insert a new blank page, but also a section break so that any page formatting you do on your new 'Next Page' will not affect its previous page and visa versa!)

Margins – 1st be sure you're in Print Layout *view* (View>Print Layout. You have to be in this view to see your margins). To change all 4 margins of your document: File>Page Setup, /Margins *tab* (When finished setting your margins /Default (b) to have these settings be permanent on all new documents & /ok).

Vertical Text Align – To align you text vertically, like you could vertically center the your Title on your Title Page: File>Page Setup, /Layout *tab* & /Vertical Alignment *arrow*>Center, /ok.

Print Preview – Before printing it's usually best to get an overview of your layout: On *toolbar* /Print Preview (b).

Print – To view print options before you print either: File>Print or Ctrl+P & select your options & /ok to print.

Styles & Formatting – You can create, apply, modify or view styles to your text in the Styles & Formatting Task Pane or TP: on *toolbar* /Styles & Formatting (b)…

    Create: You can create your own style that includes not only a font format but also paragraph too: TP /New Style (b) & type the name of your new style, then select your font & its size, then /Format (b)>Paragraph & set "Spacing" After>**6** pt, /Line & Page Breaks *tab* & check Keep Lines Together (This will prevent a page break within a paragraph. You could find this out too if you / the ? mark up in the right corner of Line & Page Breaks *tab* & then / on "Keep Lines Together"), /ok, /ok.

    Apply: 1st select some text you'd like a preformatted style applied to, then in TP / Show *arrow*>Available styles, & / on a style like Heading 2

    Modify: If a style, like Heading 1, has too large a font you can modify it to a smaller size: TP hover your pointer over Heading 1 & until you see its *arrow* then / it>Modify, /Format (b)> Font & select size 12, /ok (Note: if any Heading 1 style already applied in your documents ALL of them will be updated to size 12).

    View: You can view what styles are applied to what text, but 1st you have to be in Normal View to see styles: View>Normal, then Tools>Options, /View *tab*, & in Style Area Width *field* type **.8**, /ok. You may want to change your view back to Print Layout: View>Print Layout.

Template – An original document you don't want changed, but the original can be duplicated whereas any modifications apply to the duplicates & not the original (or Template): 1st create your document you want to save as your Template & *Menu* File>Save As, /Save as type *arrow*>Document Template.dot, & type name of your Template, /Save. To extract a copy of your Template, File>New, & //on your Template.

    Delete: *Menu* File>New, /once on your Template & /Delete, /ok.

Organizer – Is used to copy Macros and Styles to other Documents or to the Global Template (If you copy to Global Template, then every time you open Word you will have you new Style): Tools>Templates and Add-Ins, /Organizer & select your style or macro from 1 little *window* & /Copy (b) to add your Macro or Style over to the Global Template *window*.

Text to Table – You can convert all those tabs in to cells that will make up a table: 1st select all the text you want in a table & Table>Convert>Text to Table, # of columns>the # of tabs you have in each row of your selected text (make sure all rows have same # of tabs or may look funny), under Separate Text At, select "Tabs", /ok

Mouse click / Button (b)	To > Enter or Return®	Task Pane TP:

<u>Table to Text</u>: OR have you ever copied a web page into your Word document & found that they're in table format, but you'd like it in normal text? 1ˢᵗ select the WHOLE table & to do that / in any cell of the table & on *menu* Table>Convert>Table to Text, select 'Tabs' (you can select other options, but tabs works most of the time), /ok.

Merge Table Cells – To merge cells in your table: Select your cells & Right / them>Merge. <u>Split Cells</u>: To split cells (merged or not): Right / cell>Split & select # of columns & or rows to have after it's split, /ok.

Excel to Table – To insert Excel into your document as a table: In your Word document *menu* File>Open, /Files of type *arrow*>Microsoft Excel Worksheet & then browse to find your Excel *file* & //it, /ok (Note: if you want this Table in another document: / in any cell in the table & *menu* Table> Select>Table, *toolbar* /Cut (b) & open the document you want your table in Ctrl+V (paste). >Sort: To sort your table: / in any cell of table & Table>Sort, select your options to sort by & whether you have a header row (if yes make sure that option is selected or it will sort your headings too), /ok.

Table Formulas – You can add up columns or rows in your table: 1ˢᵗ find a row you'd like to add & / in a blank cell at the end of that row: Table>Formula, (note: sum (left), meaning all #s to left will be summed), /Number Format *arrow*>3ʳᵈ choice in list, /ok. If you have more rows, after you've done the 1ˢᵗ row, / in the next row's Total cell & press F4 (F4 will apply the *most recent* formatting & if the last thing you did was inserting a formula, it will copy that into the cell). <u>Update Totals</u>: The formula's Total doesn't update automatically, but Right / the Total>Update or / in Total cell & press F9.

Table Chart – To create a chart based on your table: Select your table Table, Insert>Picture, /Chart. Your chart will show along with its Datasheet…
<u>Datasheet</u>: The #s you change in your Datasheet will update in your chart…
   <u>Close</u>: WARNING! Don't close your Datasheet by clicking the X (Close (b)), but instead / off in a blank area of your document…
   <u>Open</u>: To open your Datasheet //on your chart. If you //on the chart & your Datasheet doesn't pull up then / off in a blank area of your document, Right / your Chart> Chart Object>Open; when you're finished File>Exit & Return to… and now you can // on your Chart & it will pull up Datasheet!
   <u>Hide Data</u>: In the Datasheet you can hide certain parts of your Data from displaying in the chart: >//Column Header (A, B, C etc.) or a Row Header (1, 2, 3 etc.), & it will make that row or column in your Datasheet faded & also hide that data from displaying in your chart. To unhide: //those Column or Row Headers again.

Link Excel – You can insert Excel as a table, & any changes you make in your Excel Spreadsheet will automatically update: On a blank page Insert>Object, /Create From File *tab*, check Link To File *box*, /Browse & find your Excel file & //it, /ok (Excel file is now inserted as a linked table). To make changes, Right / *table*>Link Worksheet Object>Edit Link, (Excel opens up & you can make changes that will automatically update the linked table in your document).

Clip Art – Word has a collection of pictures you can use in your document: Insert>Picture>Clip Art & in <u>TP</u> /Clip Organize *link*, Expand the 'Office Collections' *folder* & pick a category *folder*; and in the folder Right / your picture>Copy & close the 'Organizer' & Right / a place on your document>Paste.

<u>Washout</u>: AKA Watermark. To washout a picture or Clip Art so you can later put it behind some text as faded background: Select your picture or Clip Art & bring up Picture *toolbar* & /Image Control (b)>Washout

<u>Clips Online</u>: You can download more Clip Art from Microsoft's website: <u>TP</u> /Clips Online *link*, >Web page opens up, type in your keywords to search for Art & /Go (b), browse & check all art you want & then in upper-left corner /Down 2 clips (or 3 or 4, depends on how many you've selected), & follow instructions.

Drawing Toolbar – This toolbar is used to create & edit pictures & shapes

<u>AutoShape</u>: To create your own shapes & color them: Bring up Drawing *toolbar*, View>Toolbars>Drawing & on *toolbar* /AutoShapes (b)>Basic Shapes>any you like (Once you select it the 'Drawing Canvas,' new to Word XP pops up to help draw your object), Hold Shift (using SHIFT helps draw your shape proportionally) & /in upper-left corner of drawing canvas & drag diagonally down & right to create your shape. Move your pointer over upper or lower-right corner handle of canvas & / & drag your canvas smaller to fit around your AutoShape. / in a blank area off your canvas when your finished.

<u>Text Wrapping</u>: This feature allows text to wrap around your shapes, or send your shapes behind your text like a Watermark: //your picture or shape, /Layout *tab* & select: Behind Text. (Note: once your object is behind text the only way you can select that object again is on the Drawing *toolbar* / the White *arrow* (the Select Objects), and then you can //your object, /Layout *tab* & select 'In front of text,' /ok

<u>Order</u>: If some objects are hiding behind others and you want them in front: Select object you want in front & Right / it>Order>Bring to Front.

<u>Transparency</u>: Washout doesn't work on Shapes drawn from the Drawing *toolbar*, >but if you want to fade your shapes //the shape, /Colors & Lines *tab*, type in your % in the 'Transparency' *box*, /ok

<u>Color</u>: To color shapes: Select your shape & on Drawing *toolbar* /Fill (b) *arrow*>a color or 'Fill Effects' (to apply a Picture, 2 colors or Gradient, Texture or Pattern to your shape)

<u>Turn off Canvas</u>: If you would rather create Shapes directly on your document & not use the Canvas: >Tools>Options, /General *tab* & uncheck 'Automatically create drawing canvas when inserting AutoShapes,' /ok

<u>WordArt</u>: To add more visual pizzazz to a word on your document: Select your word & on Drawing *toolbar* /WordArt (b) & select a format, /ok, change its size, /ok.

<u>3-D Toolbar</u>: To add 3-D effects to any of your shapes: On Drawing *toolbar*, the last (b), >/3-D Style (b)>3-D. Be sure to select your shape 1st & then you can use 3-D Setting *toolbar* to manipulate it.

Mouse click /	To >	Task Pane <u>TP</u>:
Button (b)	Enter or Return®	

Org. Chart: To create a quick Organization Chart (for more Dynamic Charts use Microsoft's Visio): On Drawing *toolbar* /Insert Diagram & Organization Chart (b), /ok. /in a shape & type in you're the President. To add Assistants or Subordinates, select the employee's shape & on Org *toolbar* /Insert Shape *arrow*>Assistant. AutoLayout: To have complete control in sizing & moving your shapes around in your Org chart you'll have to turn off AutoLayout: On Org. *toolbar* /Layout (b)>AutoLayout.

Section Breaks – Section breaks are used to divide or section off a page(s) layout. The most popular Section Break types are…

Continuous: This type is for cutting up one page into section(s): For example, to section off a 1 page document into 3 sections where Section 1 is a single column text, Section 2 is a double column text (like a newsletter), & Section 3 is single column text.

1. Place your cursor on your page in a blank line just below your Title, but above your news story text: Insert>Break, /Continuous, /ok. (The easiest way to tell if your Section Break has been applied, on *toolbar* /Show Hide (b) to seek the code. /it again to Hide those codes)

2. Next, towards the bottom of that page place your cursor in a blank line just below your news story text but above your closing title (Closing title could be: *For More Info please call…*): & Insert>Break, /Continuous & /ok

3. Now that you've sectioned your news story text off from the top & bottom of your opening & closing titles: / any where in your news story text (Note in lower-left of your window the Statistics bar. This shows 'Sec 2,' meaning you're currently in Section 2) & Format>Columns, /Two, /ok (Note: Your top & bottom titles are in 1 column, and only your news text that you've sectioned off is in 2 columns)

Next Page: To create a new page that is sectioned off from the other pages in your document so that any Margin adjustments or additional columns apply to that new page: Insert>Break, /Next Page, /ok. For example, in a 5 page document you add a new page by inserting a Next Page *break* (Not only did this create a blank new page, but it also sectioned that page off from previous pages). If you change page 6's margins or add 2 or more columns, it won't apply to the previous pages. BUT the layout from page 6 will continue to affect ALL new pages, that is unless you continue to add your new pages by inserting a Next Page *break*.

Columns – Once you have your page(s) appropriately sectioned off you can turn your text from Word's default 1 column page into 2 or more: Format>Columns & select the # you'd like, /ok

Breaks: A Column Break prevents text from overflowing from 1 column into another: Put your cursor at the bottom of your 1st column & Insert>Break, /Column Break, /ok (repeat this for each column so when you add text into anyone of these columns & hit ® it doesn't affect the text in the adjacent column by rolling over your hard return & pushing that column's text down too).

Envelopes – To print address on an envelope: Tools>Letters & Mailings>Envelopes & Labels, type in a Delivery address (or /Insert Address (b) if you want to use your Contacts in Outlook),

& type in a Return; /on the 'Feed' *picture* & select how you'll feed your envelope into the printer, /ok, then /Options (b) & check the 1ˢᵗ two boxes:

✓ <u>Delivery Point Barcode</u>: This prints your zip code into barcode that will process your mail a lot faster, especially if you can get the additional 4 digit zip code i.e. 84121-4452

✓ <u>FIM</u> (Facing Identification Mark): Tells the Post Office's machine which is the front-side of your letter & helps process your mail faster too…

/ok, put your envelope in the printer & /Print (b).

Labels – To print address on labels: Tools>Letter & Mailings>Envelopes, /Labels *tab*, & type in your **Home Address** (For Blank labels to fill your own address type in nothing), /Options (b) & select a *label*, /ok /New Document (b)…

Mail Merge – To create a Form Letter & to address this letter to 20 different people where the only parts that change on the letter is the names & addresses etc. Before starting make sure you have

✓ A Form Letter and…

✓ A database like Excel that holds the names & address or other info that you'll later merge in certain parts on your Form Letter

Be in a new document & Tools>Letters & Mailings>Mail Merge & in <u>TP</u> 'Letters' is selected,

2. /Next, select 'Start from existing doc. & /Open & find your Form Letter & //it

3. /Next, Select 'Recipients' ('Use an existing list' is selected) /Browse *link* & look for your Excel database & //it, /ok, & at this point in your <u>TP</u> you can do a few things with your Database:

✓ <u>Sort</u>: /Last Name column *heading* (not its *arrow*, to sort ascending or descending) and or…

✓ <u>Filter</u> /State's *arrow*>Advanced: /Field *arrow*>State & in 'Compare to' *field* type **UT** (Only those clients in your Excel Database from Utah will be included in your Form Letter), /ok…

4. /Next, here you can edit your letter, Insert>Date & Time & select on, /ok & hit ® 5 times. <u>TP</u>, /Address Block, & select one, /ok®® <u>TP</u>, /Greeting Line, & select one, /ok®®

5. /Next, <u>TP</u> >> (b) to toggle through the recipients of your Form Letter (As you toggle through your recipients you can 'Exclude' some here)

6. /Next, /Save as type in a name <u>TP</u>, /Print *link* (you can choose what to what record), /Cancel. /Edit Individual Letters *link*, /ok (Your Form Letter is broken & address to each individual & you can customize each letter a little more personably…

<u>Labels</u>: To create Mailing Labels from your Excel Database: Be in a new document:

>Tools>Letters & Mailings>Mail Merge & in <u>TP</u> select Labels,

2. /Next, (Change Doc. Layout is selected), /Label Options *link* & select one, /ok

3. /Next, (Use An Existing Doc. is selected), /Browse *link* & find your Excel Database & //it, /ok, & at this point in your <u>TP</u> you can do a few things with your Database:

Mouse click / Button (b)	To > Enter or Return®	Task Pane <u>TP</u>:

✓ <u>Sort</u> – /Last Name column *heading* (not its *arrow*, to sort ascending or descending) and or…

✓ <u>Filter</u> – /State's *arrow*>Advanced: /Field *arrow*>State & in 'Compare to' *field* type **UT** (Only those clients in your Excel Database from Utah will be included in your labels), /ok…

4. /Next, /Address block, /ok. In TP, under Replicate Labels /Update All Labels (b)

5. /Next,

6. /Next, TP /Print & note you can choose what to what record to print to, /Cancel & Save.

Edit with Comments – If there's a doc being sent around the office asking everyone to add their comments & you receive it; with your mouse hover over comments to see who they're from. To leave your own: Select the text you'd like to comment on & Insert>Comment & type. If you hover over your comment & it reads "From Computer 12," you can change that so it reads *Your Name*: Tools>Options, /User Information *tab* & change the ID, /ok

<u>Quick Review</u>: To review all the Comments made on a doc: On Reviewing *toolbar* / Reviewing Pane (b), & at bottom of your screen will be a Review *window*.

Compare Documents – Instead of wasting time looking back & forth between docs that are very similar do the following: Open your 1ˢᵗ Doc & Tools>Compare & Merge Documents & check Legal Blackline, (by checking this the (b) changes from "Merge" to "Compare"), & brows & find your 2ⁿᵈ doc to compare & //it (Note: New text is <u>underlined</u> & vertical line in left margin indicates changed lines). To accept the changes: on Reviewing *toolbar* /the *arrow* to the right of Accept Change's (b)>Accept All Changes. (Note: you can now save these changes as a 3ʳᵈ new doc without affecting your 1ˢᵗ & 2ⁿᵈ docs).

Merge Documents – To merge one document into another & view changes: Open doc 1 & Tools> Compare & Merge Documents, Remove check from "Legal Blackline" if necessary & doc 2 (Note: all changes will be merged into doc 2).

Template – An original document you don't want changed, but the original can be duplicated; whereas any modifications are made upon the duplicates without affecting the original (Template): Create your document you want to save as a template & *Menu* File>Save As, /Save as type *arrow*> Document Template.dot, & type name of your template, /Save. To extract a copy of your original, File>New, & //on your Template.

Invisible Table – To create a table that provides layout structure to your document without the borders: Right table>Borders & Shading & remove borders, /Borders tab & under 'Setting' select 'None.'

Form Fields – To create a form that can be interactive for the user to fill out electronically or on paper. Note: when finished with the following fields: Fill-in, Text Form Field, Check Box, Drop-Down Form Field you'll have to: /Lock (b) on Forms *toolbar* before saving & executing them (Locking the form allows the user to only interact with the Form's *fields*, BUT you must unlock it if you want to do any editing)

<u>Fill-in</u>: To open a template as a document that prompts you with questions to answer, & those answers automatically fill in parts of your doc. For example, if you want to open a Memo

prompts you to type in the name of who the memo is addressed to & put that name in a specified part of your doc: After you've created your memo, / next to your addressee's 'To' *field* & Insert>Field, in Categories select "All" & in Fields select "Fill-in" & in Prompt type **Enter Name(s) Memo is addressed to** & then check Default Response To Prompt *box* & type its prompt *field* **Names**, /ok, /ok. (Note: Alt+F9 to toggle Fill-in codes on or off). Then save your doc as a template. To execute your prompts or Fill-in *fields*: File>New & open up your template as a doc & type in appropriate text, /ok.

Text Form Field: To insert these fields into the doc that when locked & you hit the Tab key it takes you to only these fields where you can quickly type in text pertinent to that field. For example, you can place a field like comments at the bottom of you doc: Place your cursor at the bottom of your doc next to the text "Comments" & on Forms *toolbar*, /ab| (Text Form Field) a gray field is inserted on your doc, // it to display its properties, & in "Default Text" type "Comments", /ok

Check Box: To add a box for you to check off with your mouse or with a pen on paper. For example, say I want to do a survey on household income to help with my marketing: On Forms *toolbar* /Check Box (b) & type "0-$29,000" & hit ® & /Check Box (b) again & type "$30,000-$5,000" i.e. ☐ $0 - $29,000 ☐ $30,000 - $45,000 (Note: Check Boxes & other Form *fields* are shaded in gray only to help the creating process when editing your Form, and you can turn the gray off on Form *toolbar* with a / on Form Field Shading (b))

Drop-Down: To insert a field that when you / on drops-down & displays other choices: On Forms *toolbar* /Drop-Down (b), & on you doc // the gray *field* & type "Disneyland trip" >& hit ® then type "$3,000 cash" & hit ® & type "Mope" & hit ®, /ok

Protect – To protect your doc with a password. Anybody can view your doc, but they can't make any changes to it. Tools>Protect Document & enter a password, enter it again. To remove password, follow same steps & delete the password.

Macros – A macro is a recorded series of steps that you can run in a blink of a second. For example, instead of retyping your closings on variousdocuments, record those steps & have the Macro run it for you...

Record: On Stat bar (located at bottom of window //REC (to start recording, a *mini toolbar* pops-up) & type in a name for your Macro, /ok OR...

Shortcut Keys: If you'd like to run your macro with shortcut keys then / Keyboard (b) >& use a combination of Ctrl, Alt or Shift *keys* for your macro i.e. hold down the Alt *key* & press Q, Alt+Q is added & the little box will say "unassigned," so it's okay to use, /Assign (b), /Close (b)... OR...

Button: You can /Toolbars (b) to add a to your *toolbar*, & under 'Commands' / & drag your (b) in-between some other (b)s up onto your *toolbar* above. On *toolbar* Right / your new (b)>Name & delete the extra text & leave your original name the & in the same menu go>Change Button Image>Choose a (b), /Close (b) & continue with recording... Now on

Mouse click /	To >	Task Pane TP:
Button (b)	Enter or Return®	

your doc type "Your Full Name" ® "Your Title" ® "Your email address" ® & "Phone #" ® & on Mini *toolbar* /Stop (b) (If you accidentally closed the Mini *toolbar*, Bring up the Visual Basic *toolbar* & /its Stop (b)).

Run: – To run or execute your recorded Macro on any doc: Tools>Macro>Macros & select the name of your macro & /Run (b) or use your shortcut keys Alt+Q

Edit: Editing a macro requires you understand the programming language called Visual Basic or 'VB.' But you can still edit & try to figure out some common sense editing techniques i.e. color = blue, you could change that statement to read color = red: Tools>Macro> Macros, select your macro & /Edit (b)

Delete: To delete a Macro: Tools>Macro>Macros, select macro, /Delete (b).

Bookmark – To insert Bookmarks throughout the doc as a quick reference to quickly go to. For example, if you had a 20 page doc & you wanted to keep referring to a particular paragraph (Let's say a paragraph about Doughnuts!) of your doc, but you can't because that reference shifts to a different page every time you add or delete large amounts of text in your doc…

Insert: Simply select the text you want marked i.e. Doughnut's paragraph & Insert> Bookmark & type in a name & /Add (b) (Note: if won't let you /Add then you got to get rid of any spaces in your name & or odd characters).

Go To: To quickly go to a bookmark already inserted in your doc: Press F5 & under 'Go to what' select Bookmark, then either type in the name of your Bookmark or /*arrow*>that name

View: To see your Bookmarks (as brackets around bookmarked text or as I beams next to unselected text) Tools>Options, /View *tab*, under Show check 'Bookmarks'

Delete: To delete bookmarks: Insert>Bookmark, select your mark & /Delete (b).

Footnote – Whenever want to make a reference or give credit to the text on a page you can insert a note at the foot of that page. For example, you've entered a sentence on a page that for some readers may not be clear; so insert a note at the bottom of that page referring to that sentence: / at the end of the sentence & Insert>Reference>Footnote, /Insert. You're automatically sent to the foot of the page to type in your explanation next to a reference #. (Note: After you type in your info, //its reference # & it will automatically take you back on the main page to the # being referenced to; then you can hover over that sentence's reference # & you'll see a pop-up of the text you've just entered in below as the footnote, & you can also // this # as well to go back to the bottom of the page).

Endnote – Is just like the Footnote only the reference is inserted on the very last page of your doc, not the bottom: Insert>Reference>Footnote, select Endnote, /Insert (This will automatically insert & take you to the blank page at the end of your doc with a reference #).

Caption – Used to insert text below a picture or graph & #'s the captions in order too: /in blank line below a picture or chart & Insert>Reference>Caption & type & to the Figure 1 type ":Your Caption", /ok

Cross-Reference: Used to create a cross reference between the text in your doc that are refereeing to the object(s). For example, you've already inserted a few captions i.e. Figure 1, Figure 2 etc.: Now type the text *Please see*, & / to right of 'Please see' & Insert>Ref>Cross-ref, /Ref Type *arrow*>Figure 1, /Ref To *arrow*>Entire Caption, /Insert & /Close (Note: if you hover over the shaded text it will prompt you to Ctrl+/ to go to Figure 1 instantly!)

Book Fold – Used to set your doc up to print out like a book, pamphlet, newsletter etc.: File>Page Setup /Multiple Pages *arrow*>Book Fold (Note: changes wording of right & left margins>inside & out)

> Gutter: Adds extra room for inside margins on both left & right pages for binding. If you want extra room for your inside outside margins for book binding: File>Page Setup & type amount of desired Gutter in Gutter *field* i.e. .5, /ok

> Odd/Even Headers: Like a book you can have titles for all the odd pages ("Chapter #") different or separate from all the even pages ("Chapter Name"): File>Page Setup, /Layout *tab* & check 'Diff Odd & Even' & 'Diff 1st Page', /ok. View>HeaderFooter & scroll>header of 2nd page & type "Your Name" (as the author), then scroll down to page 3's header & type "Name of your Book." File>Print Preview to see what it will print out like

> Print Book: Once you have your Multiple Pages set up>Book Fold it's important to print correctly too. >Though every printer is different here's a general way to print your book: 1st be sure to Insert page #s (Insert>Page #s) & then Ctrl+P, /Properties (b) & /Features or Advanced *tab* & look for "Two Sided Printing" & check it & uncheck "Automatic" (On some print jobs if you don't uncheck this the printer won't prompt you on your computer after it has printed one side of your book to FLIP printed pages over & insert them back into your printer so it can print the other side), & /ok, /ok.

Table of Contents (TOC) – An outline of Headings on a page with their page #s that those headings are referencing throughout the doc…**Headings –** For example, Word looks for certain styles called 'Headings' & if you had the name of each chapter in your document in a 'Heading 1' style, then Word can detect & copy all Heading 1's in your doc & put it on page 1 with their corresponding page #s, like indexing, but this will be a TOC. Sub-headings can also be added to TOC if the correct Heading style is added to it i.e. 'Heading 2' for subs & 'Heading 3' for sub-subs etc. To manually assign Heading styles: Select your headings to be on your doc & on toolbar /Style arrow>Heading 1, 2 or 3. After you have your Headings applied throughout your doc on your desired text to be copied into TOC, then create a blank page at beginning of your doc to place your TOC: *Menu* Insert>Index & Tables, /Table of Contents *tab* & set 'Show levels'>Level 1 to see only Heading 1's in your TOC or >Level 2 to see Headings 1 & sub Heading 2, & Format *arrow*>Formal, /ok

> Update TOC: After you've created your TOC & added or removed any chapters or text with their corresponding 'Headings' from your doc & you want that reflected in your TOC: Right / your TOC>Update & select 'Update Entire Table,' /ok

>> Level Change: To change your TOC's levels of Headings from reading only Heading 1>include (or not include) Heading's 2 & 3: / somewhere in the middle of your TOC & Insert>Index & Tables, /Table of Contents *tab* & set 'Show levels'>Level 3, /ok.

Index – is an alphabetical guide to concepts in your doc appearing at end of a doc with referencing page #s for each key concept. First create a Concordance file…

Mouse click /	To >	Task Pane TP:
Button (b)	Enter or Return®	

<u>Concordance File</u>: is a file containing a two column table with specific words entered in the 1st column you want Word to search for & in 2nd column the words you want displayed in your Index after:

1. <u>Create</u> a new blank doc & insert a 2 column table with several rows according to the # of key words you'd like to appear in your index.
2. <u>Left Column</u> type in all the key words you want Word to look up or search & find in your doc that want your index inserted in.
3. <u>Right Column</u> type in all the words you want displayed in your Index. For example:

Fat	Fat
Health	Health
Chicken	Meats

(Note: in 1st column, 1st cell Word will lookup 'Fat' & replace that with the adjacent cell's text in column 2, 'Fat'. Also note, although it would make sense to have all words Word looked up replaced with the exact same words, sometimes a synonym is more appropriate. In other words, you don't have to have the same text in 2nd column)

Now save your file with 'Concordance' in the name & close it…

<u>Index Marking</u>: After you've created your Concordance file, open the doc that has the related key words of your concordance file. The next step is to mark the doc you've just opened with the key words in your concordance file: Insert>Ref>Index, /Index *tab* & /AutoMark (b) & find & //your concordance.doc. (On *toolbar* /Show Hide Codes (b) to see the index codes throughout your doc)

<u>Indexing</u>: Last, but not least, once the codes have been marked its time to insert the index itself: Go the the last blank page of your doc & Insert>Ref>Index, /Index *tab*, /Formats>Modern, Check 'Right align page #s', /Tab leader *arrow*>Dots, /ok.

Master Document – holds text & links to related subdocuments. What you change in the Master will be updated in the subsdocs & what is changed in subdocs will be updated in the Master. This is great for multiple projects (docs) others are working on that need to be tracked in the Master:

1. Make sure you have all the subdocs created & ready to be inserted into the Master doc.
2. In a new blank doc (this will be saved as your Master), View>Outline View, on Outlining *toolbar* /Insert Subdocument (b) & browse & find your 1st subdoc & //it to insert it, and insert the rest of your subdocs (Note: the whole text of each doc inserted is fully displayed…)
3. Save you Master doc
4. On Outlining *toolbar*, /Collapse Subdoc (b) (Note: This creates a hyperlink when you hold Ctrl & / on will open up subdoc in a separate *window*). To Expand the link on *toolbar* /Expand doc (b)

Versions – To save the current state of your doc without saving it as new file or overwriting the original. The versions are always stored in the original doc. Open a doc you want to create a different version of: Make some changes in the doc & File>Versions, /Save Now & type in a name of this 1ˢᵗ version, /ok (Save as many different versions as you like). To open a previously saved version: Open the original doc the versions were saved in & File>Versions, select a version & /Open.

Password – Nobody can open & read your doc without a password: For example, you can put a password on a sensitive doc when you email it since emails are not always 100% safe: Tools>Options, /Security *tab* & in Password to Open type a password (Note: you can type a password in 'password to modify' box; which means that anybody can open it, but they can't modify it without the password), /ok & type your password again to confirm, /ok. Next time when someone opens the doc they have to type in a password.
  <u>Remove</u>: To remove password: Tools>Options, /Security *tab* & delete password whatever is in the 'password to open' *box.*

Track Protect – to track any changes anyone makes to doc who hasn't entered the password: Tools>Protect & type a password, /ok & type the same password in again to confirm, & save. Now any changes made in doc will show in red. To make changes without Word tracking it: Tools>Protect & type in the correct password.

Track Changes – to track changes made to your doc without a password. This can be helpful if someone sends you a letter for you to edit, but would like to see those changes you've made so they can accept or reject them without retyping their letter: Tools>Track Changes (will start tracking any changes made from here on out). Make some changes to doc & email it back to your friend.
  <u>Accept or Reject</u>: When you get a doc that has been tracked & changed you can accept or reject changes with the Reviewing *toolbar*: Go to beginning of your doc & on Reviewing *toolbar* /Next (b) & when text is highlighted /Accept (b) or /Reject (b) on *toolbar*

Shortcuts

<u>Ctrl+A</u> – Selects *all* text in document.
<u>Ctrl+Z</u> – Undo any action
<u>Ctrl+Y</u> – Redo any undid action.
<u>Ctrl+1</u> – Single space
<u>Ctrl+2</u> – Double space
<u>Ctrl+5</u> – 1 ½ spacing
<u>Ctrl+T</u> – Hanging Indent
<u>Ctrl+S</u> – *Save*
<u>Ctrl+X</u> – *Cut*
<u>Ctrl+C</u> – *Copy*

Mouse click /	To >	Task Pane <u>TP</u>:
Button (b)	Enter or Return®	

Ctrl+V – *Paste*
Ctrl+F – *Find*
Ctrl+H – *Replace*
Ctrl+P – *Print*
Ctrl+D – *Font*
F4 – Like Format Painter, *but* it only applies *most recent* action or format performed.
Cntrl+Home – Puts cursor at the very beginning of the document.
Cntrl+End – Puts cursor at the very end of the document.
Home – Puts cursor at beginning of a line.
End – Puts cursor at end of line.

Project

Project's Schedule Date – to either schedule the project to start on a certain date or schedule a project from a finish date (from finish date means that any time added or taken away after project has been built will push the start date of the project to start earlier or later date, but must finish on finish date): Project>Project Information, /Schedule from *arrow*>Project Start or Finish Date & then /Start or Finish *arrow*>a date.

Gantt Chart – When opening Project, 1st view is called Gantt Chart consisting of 2 views: Left is table & right is Gantt Chart (But ironic both views as one are still called the Gantt Chart)

> Zoom: to see all tasks in chart: View>Zoom, /Entire Project, /ok (Zooming crunches the timescale depending upon the over all length of the project i.e. your project's default timescale views tasks day- by-day, when you zoom it could crunch it to every other day, every other 3rd day, etc.)

> Timescale: is the time bar displayed directly over the chart of the Gantt Chart view, to change the time on how you view your tasks by day or hour: //Timeline

> View Bar: a display bar to help user navigate around different views in Project. To show View Bar: >View>View Bar (Note: you can also Right / on View Bar>display common Views without having to scroll in to other views in View Bar).

Project Calendar – by default Project is based on a Standard Calendar (M-F, 8-5pm with 1hr lunch 12-1pm).

> Making changes to the standard template will apply to all new projects, or you can create a calendar based on standard where you can make changes that won't apply to all new projects. Also, if there's a conflict between resource calendars & the Project's, resource *always* takes precedence: Tools>Change Working Time, /New & /Make Copy of *arrow*>Standard (8-5 Workday with 1hr lunch) or Night Shift or 24 hour & type in name of your calendar i.e. Doughnuts, /ok Options: to increase the standard 8hr work day shift to 10 (which will apply to the whole project only, and can't be applied to individual resources): Tools>Change Working Time

> Holidays: to schedule days off on project calendar: /For *arrow*>Doughnuts *calendar*, then /on a day & over to the right select "Nonworking time.

> Working Hours: to customize certain day's working hours from the standard 9-5pm: / on a day you want to change hour from 8-5 to 9-5, & over to the right, replace the current hours with your desired hours. For mass selection, in that calendar /on the *headers* M (for Monday) & it will select *all* Mondays

> Assign Project A Calendar: To assign your project a calendar (after you created it): Project>Project Information & /Calendar *arrow*>Doughnuts *calendar*.

Mouse click /	To >	Task Pane TP:
Button (b)	Enter or Return®	

Tasks – in Gantt Chart's table enter each task you'd like to have worked on in the project in successive order. If your project is going to have phases, be sure to enter each phase (or milestone) as a task and its corresponding subtasks below…

Milestones: are high-level tasks that keep track of the total # of days of all its subtasks after they are linked together in the project i.e. Electrical as a milestone, with 2 subtasks as Air Conditioning and Heating: Select the subtasks of a task i.e. Air Conditioning & Heating & on *toolbar* /Indent (b)

Show/Hide Outline: once all the subtasks are indented an outline is created. Note the – sign left of milestone task: / on – sign to collapse subtasks under Milestones & - changes to a + ready to show again its subtasks when you / on it. OR /Show (b)>hide or view levels of your Milestone outline…

WBS Codes: or Work Breakdown Structure Codes; when checked all tasks will be assigned a # that is representative of the Outline structure i.e. All Milestones will be given a whole # (2, 3, 4) & each subtask will be assigned a decimal of that whole # (2.1, 2.2): Tools>Options, /View *tab* & check 'Show Outline Number,' /ok

Duration: to set the amount of time each task is required to be completed: The column just right of Task Name *column*, type in the duration of all subtasks: i.e. 1 (1-day), 1mo (1-month), 1wk (1-week), 1m (1-minute), 1h (1-hour) (Note: all milestones will only show the duration for the *longest* subtask, and to get a total duration for *all* subtasks the subtasks *must* be linked in a relationship)

Project Summary: a task naming the project: Tools>Options, /View *tab* & in "Outline options" >check "Show Project Summary Task," /ok. Select Task 0 & type name of your project

Recurring: a task that occurs over & over again i.e. a doughnut meeting: Select the task that you want to insert your recurring doughnut meeting before it & Insert>Recurring Task & type Doughnut Meeting, Duration type **2h**, /arrow>Every Other & check Monday, /Start *arrow*>any date, /End by *arrow*>date, /Calendar *arrow*>the Project's *calendar*, /ok

Go To: to quickly go to & view one task's bar in the Gantt Chart: Select a task in Gantt's *table* & on *toolbar* /Go To Selected Task (b)

Insert & Delete: Right / Task Row *header* #>Insert or Delete. For example: I want to insert the task 'carpet' between the task 'paint' & 'move furniture in'. Right / paint's row #>Insert.

Linking: to give flow to tasks in a project, links (or relationships) are created between all tasks in successive order. So after the links are created, for example, when one task finishes the next one will start & so on. But 1st you must link all tasks together: Select the tasks in order of completion by first selecting the 1st task & then holding the Ctrl *key* & selecting their successors until all task are highlighted… (WARNING! When using the Ctrl key you have to highlight each task in order. In other words, if you select task 1 & Ctrl+/ task 8 & then task 2 the relationship will be… when task 1 is completed start task 8 & then task 2. If you made a mistake in the order of highlighting tasks, / off in a blank area to deselect & start over) …& then on *toolbar* /Link (b) (Note: The Gantt Chart shows blue bars, and when one task is done a blue arrow points to the next task to start, & this default is called the Finish-Start relationship – when this task finishes, the next one can start)

Break Links: Select tasks to be broken using Ctrl *key* & on *toolbar* /Unlink (b)

<u>Relationships</u>: when a task is linked to another task it creates a default Finish-to-Start relationship so when this task if finished the next one can start. There are 4 types of relationships:

1. Start to Start
2. Finish-Finish
3. Finish-Start (Project's default)
4. Start-Finish

To help decide what type of relationships to use, fill in the blanks of the following sentence with one of the 4 types i.e. Start-Start: The predecessor must ____ before this task can ____. To change a relationship: //a task, /Predecessor *tab* & change its default Finish-to-Start by selecting it 1st & then /its *arrow*>any relationship, /ok

<u>Lag & Lead Time</u> – //on any task, /Predecessors tab & in Lag field type in a negative # of days for 'Lead' or positive # days for 'Lag'…

 <u>Lag</u>: A delay between 2 tasks; that adds waiting time

  <u>Slack Time</u>: the amount of time a task can slip before affecting other tasks & or a project's finish date. To view slackers: View>More Views, //Detail Gantt, View> Zoom & select 'Entire Project' (Note: little green lines are slack; hover other them for pop-up notes)

 <u>Lead</u>: Starting a task a day or more before its predecessor finishes; also a -30% entered in Lag field would mean the task would start after predecessor is completed 70%

<u>Notes</u>: To insert notes: Right / any task>Task Notes & type in notes, /ok. Hover your mouse over Note *icon* in Indicator *column* to get it to pop-up *tag*

<u>Constraints</u>: will limit a Project's flexibility by altering its schedule. So it should only be used when absolutely necessary and even then it's best used only on the last Task of your project to prevent original project's end date from delaying without a warning to the user: //any task you'd like to place a constraint on, /Advance *tab* & /Constraint Type *arrow*>to a constraint for the task, then /Constraint Date *arrow*>a date you want it constrained to, /ok (Note: hover over tag in indicator column to get a pop-up of the type of constraint placed upon)

 <u>Dishonor Constraint</u>: When you set a constraint the task is then rescheduled to the date of the constraint set. To prevent tasks from rescheduling to a constraint date when a constraint date is entered: Tools>Options, /Schedule *tab* & uncheck "Task will always honor their constraint dates"

<u>Deadline</u>: is a better way to identify a deadline without project altering the schedule & preventing scheduling flexibility: //any Task, /Advance *tab* & /Deadline *arrow*>any date, /ok (The deadline can be seen in gantt chart as a green down pointing arrow. The green simply is a visual marker and does nothing more. You can also hover over the green *arrow* with your mouse for a pop- up reminder)

<u>Critical</u>: is a task that must finish on time or it will bump the project's end date out. To view all critical tasks that: In Gantt Chart on *toolbar* /Gantt Chart Wizard (b), /Next, Select 'Critical Path,'/Finish, /Format It, /Exit.

Mouse click /	To >	Task Pane <u>TP</u>:
Button (b)	Enter or Return®	

Resources – Materials and or Work you use to complete a task. To create the resource list:

View>Resource Sheet...

1. <u>Type</u> *column*: 2 types of resources: work (i.e. laborers) & material (wood, shingles etc.).
2. <u>Material Label</u>: If the Resource Name is Shingles, & the type you selected is material you can give it a Material Label that the Shingles come in i.e. Bundles. So when you assign this resource to a task you assign it in Bundles i.e. 25 Bundles
3. <u>Group</u>: this field can be use to group resources i.e. for each resource assigns it a group of 'Internal' or 'External.' And then on *toolbar* /Filter *arrow* (currently displaying 'All Resources')>Group & type in External & only those resources will be displayed. To ungroup, /Filter *arrow*>All Resources.
4. <u>Max Units</u>: is for # of people. 100% equals 1 person or laborer.
5. <u>Std. Rate</u>: is rate for laborers entered hourly, monthly or yearly i.e. 14/hr, 2000/wk, 5000/mo & 55000/y
6. <u>Ovt. Rate</u>: Over Time
7. <u>Cost/Use</u>: is a one time fee.
8. <u>Accrue At</u>: defines when costs of that resource are to be paid; either Prorated (as task is completed), Start (start of a task) or Finish (when task is finished).
9. <u>Base Calendar</u>: tells what calendar is being used or to use.

<u>Calendar:</u> All work resources entered in the Resource Sheet as Work *type* Resource Sheet, Project will automatically create a Resource Calendar for it: After creating the resource, //on resource, /working time tab, & customize the working hours, or set the days off.

<u>Availability</u>: to set when the Resource is available: /General *tab* & /Available From *arrow*>any date & then /Available To *arrow*>date (Note: if you had 100% Max Units, eventually it will read 0% because this resource isn't 100 % available *all the time* that's why it's a good idea to insert a side note with it to remind you why it's not 100%), /Notes *tab* & type in a note about this resources availability, /ok (the Note icon is displayed in the left most column called the Indicator column, & when you hover over it with your mouse the note pops-up)

<u>Assigning To Tasks</u>: To assign a resource to a task: 1ˢᵗ select a task, & Window>Split (to remove: >Window>Remove Split) At the bottom, in Split Window, under 'Resource Name' / in blank *field* & then / its *arrow*>any Resource, /ok (Note: resource isn't assigned until you /ok). By default 1or 100% of the resource i.e. Electrician is assigned. To assign more than one Electrician: change 100>200 for 2 electricians, or for part-time enter 50 (50%=4hrs out of an 8hr day) & /ok (only assign as many Electricians as you have in your Resource Sheet under Max Units).

<u>Remove</u>: to remove resource from a task: in Split Window / on Resource & hit Delete *key*, /ok (Gantt Chart will display the Resource's name adjacent to task bar)

<u>Effort Driven</u>: depending upon what Task Type...

1. <u>Fixed Units</u>: (default setting) – Assigning additional resources will shorten the Resource's *Work & Task's* duration
2. <u>Fixed Duration</u>: Assigning additional won't decrease Task's duration, but instead shortens the Resource's *Work & Units*, or
3. <u>Fixed Work</u>: is *always* checked as Effort Driven. Assigning additional will divide the *Work* between all resources but the total Work for the Task is fixed; i.e. the only time you'll notice a difference is when changing the duration won't affect the Work hrs (unlike changed the Duration on Fixed Duration), but the Unit's percentage (Note Fixed Work is always Effort Driven)

…you've selected in the Split Window along with 'Effort Driven' being checked for each Task, will determine what is cut in half & shared between 2 or more work resources assigned to the task. The 'operative' word for each Task Type is *Fixed*, meaning, if you assign more than one resource to a task, what ever is *Fixed*, won't be shared or cut in half to compensate for the additional help…

- Fixed Units *example*: Let's say then that we have a task, Ditch Digging that requires 8hrs to complete. Effort Driven is checked. You assign 1 laborer to it & it's still an 8hr task, but when you assign a 2ⁿᵈ laborer to it the duration of the task is cut in half so each laborer is only required to work 4hrs. Then you assign a 3ʳᵈ laborer to it. Then the 3 laborers works 2hrs & 40 minutes each. Never will the Units, the percentage of laborers change because the task type is Fixed Units. Note: if Effort driven isn't check then all 3 laborers work 8hrs each totaling 24hrs on ditch digging.

- Fixed Duration *example*: Effort Driven is checked. The task is 16hrs long. You assign 1 laborer to it, nothing. When you assign a 2ⁿᵈ laborer, the duration of the task remains 16hrs to complete, but both laborer's Units are cut from 100% to 50%, meaning each laborer is only required to work part-time on the task. Because the task is 2-days long then each laborer will show up and work only 4hrs, totaling an 8hr day. Then come back the 2ⁿᵈ day to finish up the 16hr duration. Never will the Duration, the amount of days of a task shorten because the task type is Fixed Duration.

- Fixed Work *example*: Effort Driven is checked. The task is 16hrs long. You assign 1 laborer to it, nothing. When you assign a 2ⁿᵈ laborer, the duration of the task is cut in half to 1day to complete, and the 2 laborer are required to work 8hrs each. Never will the Work, the amount of hours of a task shorten because the task type is Fixed Work.

Not Effort Driven: for some project managers having the duration, work or units cut in half when adding additional resources isn't practical and so most turn off Effort Driven for each task. The only thing to keep in mind is when you assign more than 1 resource to a task and you want to customize the total duration or resource's units and or work without effort driven on, and you're not getting the correct results, you'll want to check the Task Type and change it accordingly.

Overallocated: if any resource is working more than the standard 8hrs a day then the resource is overallocated or stressed. One way to see if your resource is stressed is: View>Resource Usage. Any resource in red is stressed & to go to the exact stressed dates (Note: You can also contour your resources here by changing their individual work hours – more flexibility and less problems if you do your contouring when you create your project and not contour in the middle of it): View> Toolbars>Resource Management, on that *toolbar* /Go To Next Overallocation (b). You can also Right / anywhere in the yellow *grid*>Overallocation, or Remaining Availability to show those fields in comparison to the grid's default Work *field*

Solutions: by default MS Project will overallocate any resource (put it in red) if that resource is working more than 8hrs a day. Following are some solutions to manage those overallocated resources & keep them out of the red…

Overtime: assigning overtime: In Gantt Chart, Window>Split, /on task & Right / anywhere in bottom window>Resource Work & in Ovt Work *field* enter # of overtime hrs for each resource, /ok

Leveling: is a process MS Project uses of delaying or splitting tasks to avoid conflict & as a result can delay the Project's finish date: In Resource Usage *view*, select the

Mouse click /	To >	Task Pane **TP**:
Button (b)	Enter or Return®	

overallocated resource & Tools>Level Resources, /Level Now, select 'Selected Resources,' /ok. To see what tasks were bumped around: View>More Views, //Leveling Gantt (In chart *green* bars are pre-leveled & *blue* are post-leveled)

Clear: to clear leveing: Tools>Level Resources, /Clear Leveling...

Baseline – is a base that doesn't change & is used by project managers to compare to actuals and how far off the base they are like on project's duration, task's start & finish times, costs and more. Saving a baseline takes a snapshot of those fields at that time & saves them for later comparison. Project XP can save up to 10 Baselines: Tools>Tracking>Save Baseline, /ok. To clear: Tools>Tracking>Clear Baseline, select your baseline from *menu*, /ok.

Project's Stats – to view overall of your project's statistics such as current costs, baseline costs, start, finish & actuals: Project>Project Info, /Statistics (b).

Project Summary Report – View>Reports, //Overview, //Project Summary.

Importing – you can import from Excel & Access, but the import is limited to only items in projects as Tasks, Resources & Assignments...

Excel: In project File>Open, & /Files of Type *arrow*>Excel, //your Excel file, /Next, select 'New Map,'/Next, & select 1 of 3 ways of importing into your Project, /Next & select the 1 of 3 types of data your importing (the only views in Project that Project will allow Excel to import into is Tasks, Resources or Assignments), /Next & /Source worksheet name *arrow*>a worksheet in your workbook containing your data. Left *column* reveals name fields in your worksheet to be imported & in Right *column* /in corresponding *field* & /it's *arrow*>a field name you wish it imported into in Project (this requires that you know the field names in each Project's views i.e. if I created a workbook for Tasks that I want imported, it doesn't matter really what I named the title of my Tasks column in Excel like 'TNames' so long as I map it to the correct field in Project, named 'Name'), /Next, /Save Map (if you are going to do this a lot so you don't have to remap field from your Excel book>their field counterparts in Project), /Finish.

Access: follow same steps and concepts used in Importing Excel.

Export – you can export certain data from Project into Excel from its default Templates, or you can export snapshots of views into Word or as a web Page...

Excel: File>Save as, /File Type *arrow*>Excel, /Save, /Next & select 'Selected Data Option,' /Next & select 'Use Existing Map' (only if you have one previously saved or would like to use one of MS Project's templates like 'Cost Data by Task'), /Next, select 'Cost Data by Task' & you can /Finish.

Picture: take a snapshot of a view in Project & paste it into MS Word for a variety of reasons, but one could be to email off to those who want to view parts of your project but don't have Project talled on their computers: In Gantt Chart on *toolbar* /Camera (b) & select 'For Printer' (black & white so doesn't use up expensive color ink), & then open Microsoft's Word & /Paste

Web Page: to a copy of project in a universal readable format, a web page: File>Save as Web Page & type name of page & /Save, /Next, select 'Use Existing Map' & /Next, select 'Export to HTML...' & /Next, /Finish.

Tasks – for more advanced features with tasks…

> <u>Updating</u>: to keep track & update a task's progress: View>More Views, //Task Sheet, then View> Table>Tracking, Select a Task's %Comp. *field* & type in the % complete ® (Note: after % complete is entered MS Project calculates the Actual & Remaining Work & Costs for resources assigned to the task. If you want to do this manually without MS Project's help then: Tools>Options, /Calculation *tab* & uncheck 'Updating task status updates resource status'). You can then enter Actual Duration in the Act. Dur. *field* enter any # of days or Actual Cost, Work etc.® (Note: don't enter the Actuals before the current date or MS Project will calculate funny results)

> <u>Splitting</u>: splitting a task is usually done when a task has started, but then encounters a delay & hence the rest of the task is split to finish at a designated later date: In the Gantt Chart hover your mouse over the task's bar & Right / it>Split Task, then slowly move your pointer over the same task bar & find a date to break the task & / (Note: the task is now split, but only delays the 2nd half of the split task out 1-day, if the delay is more simply / & drag the 2nd half of the task out as many days delay as desired)

>> <u>Combine</u>: to combine a split task back into a whole one: Hover over the 2nd half of the task until you see a 4-way arrow & then / & drag 2nd half back to 1st half.

> <u>Rescheduling</u>: like Task Split, if a task is incomplete & will complete at a later date: Select a task marked partially completed & Tools>Tracking>Update Project & select "Reschedule uncompleted…," & /its *arrow*>any date i.e. 8/16/09 (the remainder of the task will always start the day after the set date so this one will start on 8/17/09) & select "Selected Task," /ok

> <u>Slipping</u>: to view all task that have slipped: On *toolbar* /Filter *arrow* from 'All Tasks'>Slipping Tasks. To remove Slipping Filter change Filter *arrow* back to 'All Tasks.'

Tracking Gantt – will show the baseline and compare it to the task's Actuals in the chart: View>Tracking Gantt (Note: if a baseline has been saved the chart will display it as a black bar for each task).

Progress Line – in the Gantt Chart you can view display this red line that when straight means all tasks are on target and marks the current date, but if any part of the red progress line bends back on any task, it means that task is behind schedule). Right / in Gantt Chart>Progress Lines & check "Always Display Current Progress Line" & select "At Current Date," /ok.

Tracking Toolbar – mainly used to mark of percentage of task completed: Right / any (b) up on any *toolbar*>Tracking. Select a task & /a percentage (b) on Tracking *toolbar* to mark its state of completion.

Variance Table – This table is has 2 columns that compare the current Start & Finish *fields* of each task to its Baseline & displays those in the Start & Finish Variance *columns* i.e. all negative #s in Start or

Mouse click / Button (b)	To > Enter or Return®	Task Pane <u>TP</u>:

Finish Var *column* means that the tasks are finishing early (-2 means finished 2 days early), and like wise all positive #s: View>Table>Variance.

Interim Plan – only saves or takes a snap shot of start and finish dates of each task at time saved at Interim. Up to 10 plans can be saved: Tools>Tracking>Save Baseline & select 'Save Interim,' /ok

<u>Custom Table to View Interims</u>: View>Tables>More Tables & select Baseline, /Copy (to make a copy of the Baseline view to edit) & for Name type **Interim Dates** & check "Show in menu" (this will display 'Interim Dates' Table in the Table *menu*). Delete any fields (Rows) you don't want to see in your table. Then /In a blank *row* & /its *arrow*>Start 1 (This is Interim Plan 1) ® then go to Start 1's Title *cell* & type in its name as **Interim Start 1** ® Repeat for Finish 1 & each additional saved Interim Plant i.e. Start 2, Start 3 etc. too…, /ok, /Apply.

Custom Fields – creating a custom field & inserting into any view as a custom column to help manage your data in project…

<u>Text</u>: This field is a generic one to create because anything can be entered into this field including text & #s for whatever you need an extra field for: Tools>Customize>Fields, /Type *arrow*>Text (this type is for a field that will allow you to type in any text), /Rename & type a name for your custom field, /ok, /ok. Right / any Column *header*>Insert & /Field Name *arrow*>the name of your field, /ok.

<u>Remove Column Fields</u>: to hide a column like Task Name, Right / its Column *header* 'Task Name'>Hide

<u>Flag</u>: This field is a yes or no field which again is used for anything you can thing of that is a yes or no situation for a task i.e. if certain tasks need to pass an inspection this cutom field could display a happy green face for yes or a frowning red face for no: Tools>Customize>Fields, /Type *arrow*>Flag (To check any custom field's values like this yes no Flag, /Value List (b), /in Value list *cell* & /its *arrow* to note 'Yes'/'No' Values, /Cancel), /Rename (b) & type in a name for your field, /Graphical Indicator (b)…

1. /Test Flag one *arrow*>equals & under value type **Yes**, /Image cell's *arrow*>a HAPPY FACE.
2. Repeat step one but this value type will be **No** with FROWN FACE…, /ok, /ok

Right / any column header>Insert *column* & /Field name *arrow* >Flag 1 or the name of your flag & test its fields in the column …

Hyperlinks – linking task to a document or spreadsheet: Right / any task>Hyperlink & browse to find a document & //it. (Note: the link to / on will only display itself in the Indicator *column*. Right / any Column *header*>Insert Column, /Field Name *arrow*>Indicators, /ok to insert Indicator *column* if one isn't already displayed).

Custom Reports – View>Reports, //Custom, /New, /ok, for Name *field* enter a name for your report, /Table *arrow*>any table that interest you to base the report on, /ok, / Preview…

<u>Headers & Footers</u>: Then /Page Setup, /Header *tab* & under its Center *tab* delete & or enter info that will be displayed on all pages of the report in the header's center. >/General *arrow*>Project Title & /Add (to insert Title of Project), & or /General *arrow*>Project Current Date & /Add, & or /Page Fields *arrow*>%Complete & /Add & type after the interested code type **Complete** (to label what the # inserted represents, again the %Complete), & or /Footer *tab* & under its Left *tab* & /Insert Picture (b) & browse & //any picture you'd like as part of your report & /Print Preview…

<u>Print Range</u>: some of MS Project's reports will allow you to customize the date spread to be printed: >/Close to close out of the Print Preview & back to Custom Report selection *window* & select 'Task Usage' & /Preview… (Note: many pages, but we can narrow the date to be printed…) /Print (b) & change the date to print from i.e. 10/6/09 to 11/9/09, /Preview…

Templates – you can create a basic project with tasks & outlines that can be saved as a template to copy & base future projects off of it: File>Save as, /File Type *arrow*>Template & type name of template, /Save & check what types you don't want saved into your template, /Save & File>Close. File>New & in <u>TP</u>, /General Templates *link* & //your template. Now when you save this it won't overwrite your template(.mpt) but Save as a copy of it (.mpp).

Custom Views – to create your own specialized views. There are 2 types single view & combo:

<u>Single</u>: View>More Views, /New & select 'Single View,' /ok & type a name for your view, /Screen *arrow*>a desired screen, /Table *arrow*>to a closely corresponding table of desired view, /Group *arrow*>No Group, /Filter>All Tasks, /ok, /Apply

<u>Combo View</u>: this view will display two views in a horizontal window split in Project: View>More Views, /New & select 'Combo', /ok & for Name type a name, /Top *arrow*>a view, /Bottom *arrow*>another desired view, /ok, /Apply.

Organizer – is used to copy almost anything custom made like Views, Tables, Calendars, Reports, Forms, Fields etc. into another project or onto the Global Template so your customs can be displayed in all new projects. Also, this is the place to delete them from your project as well: Tools>Organizer. Left *pane* contains Global Template & in Right your current opened Project. / the appropriate *tab* i.e. Calendars & in Right *pane* select a custom calendar you want *always* in all new projects created from here on out & /Copy (b) (it copies that selected field over to the Global Template). To share only between 2 projects & not put custom on Global Template: You must open up the 2 projects & in the Organizer's Left *pane*, at the bottom, /its *arrow* from Global> your 2nd opened project's name & then you can now copy between the 2.

Share Resources – if you have resources that many other project's will be using, the best thing to do is to create & save a Project with only the resources listed in the Resource Sheet and nothing else &

Mouse click /	To >	Task Pane <u>TP</u>:
Button (b)	Enter or Return®	

name it 'Resource Pool': After, open up a project that will be using resources from the 'Resource Pool' & from its *menu* Tools>Resource Sharing>Share Resources & select 'Use Resources,' & make sure From *field* displays 'Resource Pool,' /ok (this will then make a copy of all the resources form the 'Resource Pool' into your project's Resource Sheet).

Overallocation: when sharing resources from same pool, that pool will keep track of all resources, in that you can find out when you assign a resource if that resource is overallocated because some other project is using that resource at the same time. For example, after assigning a resource to a task in your project: View>Resource Usage & look for red resources to see if overallocated & if so there's no current way of getting MS Project to displaying how many other project's are using that resource & specifically what times, so you'll have to do the investigation yourself.

Master Project – is a project that oversees subprojects. Any changes made in subprojects will be updated in Master & any changes in the master will be updated in the subprojects as well once you. First start with a blank Project and: Insert>Project, //browse & //a project (Note: the title of the inserted subject is displayed as Task 1 with a + sign that will expand to reveal subproject's tasks). /below Task 1 (in a blank Task field) & insert your 2nd subproject & repeat steps as necessary (Also, you can link your subprojects in relationship i.e. Finish-to-Start to continue one after another in your master, just like you did your tasks).

Microsoft Project Server – after your IT person has set up the server for MS Project, then the Project Manager is able to do the next couple of steps to manage his project on the server & also how others can access that server. Note: Client Access License (CAL) is needed for each person who wants to view your project on the server.

Publish Project To Server –

1. Open your project you want on the server & Collaborate>Collaboration Options: /Collaboration using *arrow*>Microsoft Project Server. In URL type (your IT dude ought to provide this to you, something like…) **http://server/projectserver**, then /Test Connection (b)…, /ok. Under 'Identification' select "Window user account" This allows access to Project Center without having to log onto the Project Server separately, but logs on using your Window user account), /ok

2. Publish: Collaborate>Publish>Project Plan, /ok (to save after published), /ok (to publish Plan). If prompted 'make trusted site,' /yes, /ok (now snapshot of your Project is on server)

3. View Published: To check & see if project was published we can use your opened Project as a window to the Server: Collaborate>Project Center (Project Center is Home page. If you use User Names at your office to log on you'd enter them at this point), /the name of your project's *link* (to display published project plan's task), then in upper-right corner. When finished confirming, /Click Here To Close This View *link* (Note: Deleting Info from Server you will need your administrator to help you out)

Assign Resources: Next to assign resources to your tasks & publish it to server so later when the resources log on they can accept or reject assignments:
- Select a task in your project & if you haven't done so already assign resources to it. With Task 2 still selected: Collaborate>Publish>New & Changed Assignments, /ok (to save Project after published), /ok & select 'Selected Items' & uncheck 'Notify…" (if not using email), /ok (to publish), /ok

<u>Review Resources</u>: this is how your resources will log on to view their assignments:

1. The resource opens up the Internet & goes to the web address designated by the IT dude i.e. http://server/projectserver & log in (again the IT person can give them log in names & passwords).

2. Under 'Tasks' /the new task *link*, /on the assigned *task* (Note: Indicator Column for task has a *sparkling* Task *icon*; it will display only once when its first viewed), /Go To Task (b) to view task assigned.

<u>Request Progress</u>: to request progress reports from your resources on a specific task:

- In Project: select a task you want a progress report on & Collaborate> Request Project Info, /Request Progress Info for *arrow*>Selected Items, /From *arrow*>& select a date, /To *arrow*>ending date (Note: Indicator column will eventually display an 'envelope with a clock & a ? mark' indicating that your project is waiting for an update report on the task), /ok...

<u>Actual Work</u>: once the request has been made, now it's the resources turn to enter in their actual work on their end:

1. Resource goes to website & logs in, (Note: task indicator *column* has a ? mark; saying that Project Manager has requested progress info)

2. /Timesheet view *link*, then expand 'View Options,' /Date Range *arrow*>select the date, /To *arrow*>end date, /Set Dates & then Collapse 'View Options.'

3. In left pane scroll>Actual Work *field* & type actual hours work & hit ®

4. Select the *task* & /Update Selected Rows (to send update for approval from Project Manager), /ok (Note: indicator has changed from ? mark to a Sheet with a ? mark: saying its been sent but not updated in project plan)

<u>Accept Updates</u>: now the project manager is back in his project & can check and accept any updates: >Collaborate>Update Project Progress. /in Accept? *cell* to left of the *task* for the 1ˢᵗ resource, & /its *arrow*>Accept. Repeat steps for rest of *resources*. Upper-right *corner*, /Update, /ok (saves project to server), /ok, /ok. Then in upper-right *corner*, /Click Here To Close This View *link*.../Save.

Mouse click /	To >	Task Pane <u>TP:</u>
Button (b)	Enter or Return®	

About the Author

Early on in his career Kirt Kershaw received "The Distinguished Service Award" from the Jordan School District's Superintendent for successfully help graduating 24 out of 25 at-risk high school students. Realizing his creative outreach skills in communications both written and oral, he applied and got accepted in one of the nation's top journalism schools, the University of Utah.

He graduated there with a BS in Mass Communications, and an emphasis in Broadcast Journalism. His certifications include: Microsoft Office Specialists Master Instructor 2000 and XP, Microsoft Project XP and A+. He is the President of Dream Force LLC a software training company and can be contacted at 801-278-0892, or found on the Web at www. dreamforce.us.

9 781418 484569